LIVING
FOSSIL

LIVING FOSSIL

The Story of the Coelacanth

KEITH STEWART THOMSON

W. W. NORTON & COMPANY · NEW YORK · LONDON

Printed in the United States of America.

The text of this book is composed in Baskerville.
Composition and manufacturing by the Haddon Craftsmen, Inc.
Book design by David Levy

Library of Congress Cataloging-in-Publication Data
Thomson, Keith Stewart.
    Living fossil : the story of the living coelacanth / Keith Stewart
Thomson.
        p.   cm.
    1. Coelacanth.   I. Title.
    QL638.L26T46   1991
    597'.46—dc20   90-43053
ISBN 0-393-02956-5
W.W. Norton & Company, Inc.
500 Fifth Avenue, New York, N.Y. 10110
W.W. Norton & Company, Ltd.
10 Coptic Street, London WC1A 1PU

2 3 4 5 6 7 8 9 0

Contents

Preface   7

Facts   13

PART ONE   EARLY DAYS   17

   1   Courtenay-Latimer and Smith   19

   2   To the Comores   50

   3   A Living Fossil   70

   4   The Comores: Catches, Observations,
       Early Results   103

PART TWO   ANSWERS AND QUESTIONS   127

    5 Bringing the Pieces Together   129

    6 Where Do They Live?   134

    7 How They Live: Swimming and Feeding   153

    8 Physiology and Behavior   171

    9 Reproductive Biology   190

   10 Coelacanth Relationships and the Origin of
       Tetrapods   202

   11 Population Size, Conservation, and the
       Future of *Latimeria*   219

Notes   235

Index   247

Preface

Ex Africa semper aliquid novi.
There is always something new out of
Africa.
—*Pliny the Elder*

This is, simply enough, a book about a fish. But nothing interesting is ever really simple. Very soon after I became interested in the "living fossil" fish *Latimeria chalumnae,* better known as the coelacanth, I realized that I was studying a legend as well as an animal. As the years moved by, I have grown fascinated as much by the stories of the people associated with this fish as by the scientific results of their labors. The romance of discovery after its group was thought to have been extinct for eighty million years, and the fact of its possible relationship to higher vertebrates including man, have given this big ugly fish a sort of star quality.

The story of the coelacanth, perhaps the most famous species of fish in the world, has been told before, from its first discovery off South Africa in December 1938 to the latest revelations from the Comoro Islands. Some of the old stories have now been embroidered with a rich weave of invention, and new stories arise, ever more fanciful: The oil from coelacanth flesh has aphrodisiac properties; the oil causes liver cancer; the fish really lives in freshwater springs under the ocean; it is the direct ancestor of all land vertebrates (including man); it is a sort of shark; it also lives in the Mediterranean and/or the Red Sea. Some of these myths have been lost and then rediscovered (roughly at ten-year intervals). Music has been written about the coelacanth, and Ogden Nash once wrote a poem in its praise. The fish has been the subject of countless cartoons. Not a few otherwise sober scientists have floated into flights of embarrassing fancy by "old fourlegs," as J. L. B. Smith so inappropriately named it. And there are more self-appointed experts on the coelacanth than there are specimens.

In the rush to cash in on the rarity of the coelacanth, enterprising types have purchased specimens from the Comores and attempted to sell them at a large profit to museums around the world. Cashing in on the glamour, institutions have sent expeditions to the Comores to catch live specimens and have counted it a great success (at least for the press) to come back with a preserved specimen purchased from the government. Not surprisingly, others have fabricated accounts of mysterious captures or sightings of the fish and sold their stories to the press.

Up to now the remoteness of its island base and the rarity of capture of specimens have seemed to assure the survival of the species. In fact, no one, other than Comoro islanders and the trawler captain who snared the first one, has ever caught a coelacanth. Several Western expeditions equipped with the latest fancy gear have tried and failed. But now there is every reason to fear that the very attraction of the coelacanth will cause its extinction. Too many specimens are being caught by the islanders for sale overseas, either officially or on the black market. And now that the technology exists to go after the fish underwater, attempts to capture live specimens for exhibit (for enormous potential financial reward) are inevitable.

In this book I attempt to retell the story of the living coelacanth and its relatives and thereby to unravel some of the mysteries concerning its biology: How does it live? Indeed, *where* does it live? How does it reproduce? How has it managed to avoid extinction over the last eighty million years? And what are its chances now? The story of the coelacanth is also the story of a large number of fascinating people, many of whose names will never be recorded. It is a story of careful planning and blind luck, of human determination and certainly of human foibles, where experience is more useful than a hundred-thousand-dollar machine. In other words, it is not a Hollywood version of science, full of white-coated saintliness and brilliantly executed experiment, but it is pretty much a story of what science (good and bad science) is really like, especially when it involves biologists getting out in the field, soaking wet. When one strips away all the mythology and all the media "hype," the coelacanth is simply a fish, a very large, prickly-scaled, oily, sluggish, but sharp-toothed fish that, when dead, is particularly smelly. It lives in the sea "as men do a-land; the great ones eat up the little ones," as Shakespeare said.

In researching this book I have been surprised by a number of inconsistencies. In addition to the usual errors and sheer laziness of science journalism, even different versions of accounts by some participants have been inconsistent. Eyewitnesses of events have adjusted their versions over the years. Information claimed as firsthand has turned out to be secondhand after all, and usually elaborated in the transfer; a romantic overlay has been applied, especially to the role of Professor Smith; a dramatic march through the mountains turns out to have been a ride in a truck, for example. My greatest fear is that while I might clarify much of the story, the fact that we all are forced to rely on a few basic sources means that errors remain and may even have been reinforced here.

Finally, events always move quickly with respect to the coelacanth. At the time of writing (early 1990) the Comoro Islands are in political upheaval once again; a new group of play-ers has taken center stage in the scientific drama—namely, a well-supported Japanese research team that is hard at work trying to capture a live specimen.

I am indebted to a great number of friends and colleagues for their encouragement and assistance over the years in this and other projects, in particular the late A. S. Romer, Robert Griffith, James Atz, Robert Giegenbach, C. L. Smith, Bobb Schaeffer, Humphrey Greenwood, Colin Patterson, Peter Forey, John Maisey, Robert Mc. Peck, and all my students at Yale. Dr. Atz gave me access to his marvelous collection of cuttings and articles on the coelacanth, including his copy of Smith's "poster," and also arranged with the American Museum of Natural History for use of two of the photographs reproduced here. Dr. M. Bruton of the J. L. B. Smith Institute in South Africa was of great assistance in obtaining copies of important photographs in its collection pertaining to the first two captures of coelacanths. Michelle Press, managing editor of *American Scientist* magazine, kindly agreed to the use of mate-rials that had previously been published in two articles by me in *American Scientist* in 1986 and 1989. Mrs. Maureen Joubert, librarian at the South African Embassy in Washington, was extremely helpful with South African geography and biogra-phy. Dr. Richard Greenwell allowed me use of two of his photographs, and the whole community should be grateful to him for his sensitive interviews with Miss Courtenay-Latimer and Mr. Goosen in South Africa.

Bob Griffith and Jim Atz also generously read the entire manuscript, correcting as many errors as their patience could bear and improving the scientific treatment in innumerable ways.

The staffs of the Free Library of Philadelphia, Yale Univer-sity Library, and the Boston Public Library have helped me greatly, and I owe a great debt in particular to the entire staff of the Library of the Academy of Natural Sciences. I am grate-ful to my indefatigable administrative assistant, Sheryl Harris, for her patience, to Jessica Thomson for help with library research, and, as always, to Linda Price Thomson for all the drawings. A lot of my work on living and fossil fishes has been supported by the National Science Foundation, whose role in maintaining a balanced spectrum of science support is rarely completely appreciated.

Facts

Name: *Latimeria chalumnae* Smith 1939. Named for Marjorie Courtenay-Latimer, discoverer; for the Chalumna River, Cape Province, South Africa, near the estuary of which the first catch was made, and signifying that Professor J. L. B. Smith first studied and named the fish.

Vital Statistics: Length up to 180 cm and weight 95 kg; color shades of blue with pinkish white blotches, changing after death variously to purple-brown.

Date First Caught: December 22, 1938.

Where First Caught: The estuary of the Chalumna River; caught in a trawl operated by the fishing boat *Nerine* (Captain Hendrik Goosen); depth approximately 70 meters.

Subsequent Catches: None from South Africa; all from the Federal Islamic Republic of Comores (Comores, Comoro Archipelago, or Comoro Islands), a group of four small islands between the northern tip of Madagascar and the African main-

land. Second specimen caught by Ahmed Hussein Bourou and
Soha on December 20, 1952, off Domoni, Anjouan Island. The
fisherman gave it to Captain Eric Hunt, who gave it to Professor Smith. Since that date some 150 to 200 specimens have
been collected from the same island group.

Taxonomic Position: A lobe-finned fish (Sarcopterygii), member of the order Coelacanthini, a group of relatively primitive
bony fishes. Closest known relatives are coelacanth fishes of
Cretaceous age (roughly seventy million years old) from
Europe and South America. Related to the living fossil Dipnoi
(lungfishes) and, more distantly, to the ancestors of the land
vertebrates (amphibians, reptiles, birds, and mammals).

Present Conservation Status: Uncertain, probably endangered.

How do you pronounce it?

COELACANTH—"seel-uh-kanth"

PART ONE

EARLY DAYS

Courtenay-Latimer and Smith

[T]he most beautiful fish I had ever
seen was revealed.

—*M. Courtenay-Latimer*

This is the story of a fish, a particular kind of fish, steely-blue and huge, some of which are nearly six feet long and weigh 150 pounds. The first-recorded catch of this fish was scarcely glamorous. It was scooped up by a trawl dragged along the muddy bottom of the Indian Ocean, just off the mouth of the Chalumna River in southern Africa. Later it was left in a pile of sharks and other fishes under the morning sun, hardly looking like one of the most important zoological discoveries of the century. Our story might begin on the deck of that trawler before dawn on a hot December morning, in 1938. Instead, it begins twenty years before that, with a little girl staying at her grandmother's house, lying awake at night and watching the beam of a distant lighthouse touching the windows.

In December 1938 Marjorie Courtenay-Latimer was the young curator of the East London Natural History Museum in the fishing port town of East London, Cape Province, South Africa. The East London Museum is still a small museum, but in 1938 it was very small indeed. The major institution for the study of natural history in South Africa in those days was the South African Museum in Cape Town. But several regional museums had recently been established, and the East London Museum was among these, having been founded in 1930. Its shoestring budget ensured that the curator was expected to fill every role from scientist and librarian to typist. It was quite unusual for a woman to have this position, but Courtenay-Latimer seems to have had a supportive board of trustees. Given charge of building up the museum's program of exhibits and collections, starting with the usual eclectic assemblage of gifts and bequests with which all small museums begin, Courtenay-Latimer found her principal opportunities were in regional natural history. With fishing as a major local industry, she decided to concentrate on marine life, and she soon built up a network of contacts with the local fishermen, who would bring any special and unusual specimens to her for the museum.

In an autobiographical article published in 1979, Courtenay-Latimer recalled that as a child she used to visit her grandmother's house on the coast, and from there at night she would see the light from the lighthouse on Bird Island, one of a group of islands in the Indian Ocean about twenty-five miles east of Port Elizabeth.[1] Few of us are untouched by the romance of a lighthouse's rays at night, sweeping the ocean from an inaccessible and dangerous place. These islands had been the scene of a famous shipwreck, that of the *Doddington* in 1755, when twenty-three castaways were marooned for seven months until they could build a boat and escape. As a girl she longed to explore the islands. As a budding naturalist she wanted to see the colonies of seabirds and to explore the remote rocky coasts with their tide pools and abundant marine life. But it seemed that she never would—especially because she was a girl.

Many years later, after Courtenay-Latimer had been appointed curator of the new museum and threw herself into collecting the marine resources of the province, she resolved once again to visit the lonely group of islands off the coast. In 1935 she met with a Captain Patterson, the government administrator in charge of "Islands off South Africa," and persuaded him to allow her a six-month stay, to study and collect the natural history of the islands. She had to agree to be accompanied (by her mother, although in the end, greatly to her distress, her father insisted on coming, too).[2] But finally she made it to the islands and spent an idyllic but very hardworking time collecting fishes and seabirds.

While working on Bird Island, Courtenay-Latimer found that fishermen landed on the island frequently, so she got to

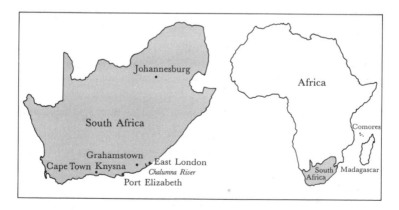

FIGURE 1   East London is in South Africa's eastern Cape Province. The Chalumna River is just to the southeast.

know them, and she collected and preserved the interesting
creatures that they brought up with their nets, lines, and traps.
One of the most sympathetic to the young woman scientist
trying to establish her museum was Captain Hendrik
("Harry") Goosen of the Irvin and Johnson Company. He operated the trawler *Nerine* along the coast, and when the six
months were up, it was Goosen who took off her fifteen crates
of materials for the museum. Over the years the captain continued to keep a special lookout for anything that might interest his friend the curator, and he also collected live material for
the East London Aquarium.

Much of what Courtenay-Latimer collected along the south-
eastern coasts of South Africa was new to science. Southern
Africa was still little explored, especially in the oceans. She
often had to send specimens away for identification and study
although there were few enough experts anywhere in Africa to
help her with the difficult cases. One person in particular, Dr.
J. L. B. Smith, a teacher of chemistry at Rhodes University
College at Grahamstown, understood her interests, for he also
spent time at sea with the commercial fishermen.

On December 22, 1938, Captain Goosen and the *Nerine*
(Smith later erroneously said it was the *Aristea*) put into East
London with a mixed catch of fishes. As usual there were
sharks, rattails, and redfishes, but the *Nerine*'s crew had also
kept a peculiar large fish, some five feet long, of a sort that
neither Goosen nor the crew had ever seen before. The deck-
hands called it a "great sea lizard" because of its odd fins,
which looked like finny legs. As usual, when they got back to
port in the morning, Goosen put to one side this unusual fish
and all the others that would not sell, in case Courtenay-Latimer should want them. The story is best continued in Marjorie
Courtenay-Latimer's own words:

> 22 December 1938 dawned a hot, shimmering summer's day.
> At 10.30 my newly installed telephone rang to say that the
> trawler *Nerine* had docked and had a number of specimens for
> me. I was busy completing the creating of a fossil reptile in a

case, and at first thought "what shall I do with fish now? So near Christmas." Then I considered I should go down and wish the men on the trawler a "Happy Christmas." So I rang for a taxi and went down to the fishing wharf. It was now 11:45 and all the men had left leaving an old Scotsman who said "Lass they have all gone but I will show you the specimens set aside by Captain Goosen." I went onto the deck of the trawler *Nerine* and there I found a pile of small sharks, spiny dogfish, rays, starfish and rat tail fish. I said to the old gentlemen "They all look much the same perhaps I won't bother with these today"; then, as I moved off, I saw a blue fin and pushing off the fish, the most beautiful fish I had ever seen was revealed. It was 5 feet long and a pale mauve blue with iridescent silver markings. "What is this?" I asked the old gentleman. "Well lass," said he, "this fish snapped at the Captain's fingers as he looked at it in the trawler net. It was trawled with a ton and a half of fish plus all these dogfish and others." "Oh," I said. "this I will definitely take to the Museum. . . ."[3]

The strange fish had been taken in about 40 fathoms (240 feet, 70 meters) as the trawler was making a long drag along an elliptical course about 5 miles offshore and 18 miles southwest of East London, near the mouth of the Chalumna River (sometimes called the Tyolomnga). This was not a commonly trawled region. The bottom is gently shelving and muddy, but just a little way farther offshore the depth falls off sharply to 600 feet or more and a region of rocks.[4] The fish had been found near the tip, the cod end, of the net, so it must have been taken early in the trawl. Even so, it had not been killed by the press of other fishes that were piled on top of it. Under such conditions most fishes would have been dead before they were brought to the surface. In fact, Goosen had made this last run to get live specimens for the aquarium but caught so much that most of the catch asphyxiated in the net and was useless. This big blue fish may well have been trapped within the net for more than four hours during the trawl. Remarkably, it had then lived for three to four more hours on the deck of the trawler after the net had been emptied. Just how lively it had been when it was taken from the net may be debated. Smith in his 1940 monograph said that it was "aggressive, snapping viciously at nearby hands," which is much more dramatic than what the deckhand told Courtenay-Latimer.[5] By 1956 this had become: "When [Goosen] touched the body, it heaved itself up suddenly, snapping its jaws viciously and had nearly caught his hand in its formidable fang-lined mouth."[6] In an interview fifty years later Captain Goosen said that it simply "snapped its jaw shut" when he tried to look at its teeth.[7] This fish was evidently alive, but only barely.

Naturally, it took a little persuasion to get her taxidriver to take on such a big, smelly fish, but Courtenay-Latimer and her assistant Enoch eventually got it to the museum.

What could this new fish be? Latimer had by that time become quite familiar with both the fishes of the inshore regions and the fishes that the trawlermen caught in deeper water. But this was different from anything she had ever seen or heard of. The fish measured 54 inches (150 centimeters) overall and weighed 127 pounds (57.5 kilograms). As it lay in the laboratory, its color continued to change. It had been a bright steely blue when caught, with pale, irregular blotches. It steadily changed to a gray-black, with the blotches persisting. It had only a few teeth in its mouth, and the whole body, including the fins, was covered with large, hard, bony scales with sharp, prickly spines. The fins were most peculiar. The tail fin was a very unusual shape, being quite symmetrical and having a short extra middle lobe that stuck straight out. The fish had two dorsal fins, whereas most other fishes (except sharks) have but one. The more anterior of the two dorsal fins was fanlike, much like that of a large grouper, but the other was stubby with a scale-covered base and a terminal fan of rays. The second dorsal fin was exactly matched by an anal fin of the same construction. All this is quite unlike anything seen in any other fish. But the most unusual feature was the paired fins. They also consisted of a stout, stubby stem to which a fan of fin rays was attached, and they flopped in every direction, for all the world like legs.

Courtenay-Latimer remembered that the lungfishes (particularly the Australian lungfish, *Neoceratodus forsteri*) have paired fins something like this but that all living lungfishes are found in freshwater and are distinctly different. This was not a lungfish. She knew from her reading about fossil fishes that big armored scales were something found in the very primitive, extinct fishes that paleontologists sometimes called ganoid fishes, to which the lungfishes were related. Whatever this fish was, it was probably important; she would need help. First she needed to preserve it. She was adept at preserving smaller fishes in jars of alcohol or formalin and had mounted smaller fishes for display. But this huge fish was beyond her capacity, and because of the heat, she would need help in that direction immediately.

She took a set of photographs and gave them for developing to a Mr. Kirsten who often did photographic work for the museum. The chairman of the board of the museum, a local doctor, J. Bruce-Bays, "a very sarcastic old gentlemen," came by and declared that the fish was only a rock cod, but Courtenay-Latimer knew better.[8] Then she tried both a cold storage warehouse and the mortuary of the local hospital for assistance in keeping the fish intact while she decided what to do with it. But neither would give even a temporary home to such a large and smelly animal. So she asked another trustee of the museum, W. E. Sargent, for the loan of a hand truck (evidently the museum was too poor to have its own), and she and Enoch trundled the fish through the streets of East London to the taxidermist on whom Courtenay-Latimer relied for bigger jobs—Robert Center—even though she knew he was "not good at fish mounting." Center at once agreed to help, and first they tried covering it with cloths soaked in a small quantity of formalin obtained from the pharmacy and wrapped the whole thing in paper.

For expert advice about the fish, Courtenay-Latimer knew where to turn. First she tried to telephone, and then she sent off a letter with a sketch of the fish, to the man who was then the only active ichthyologist in South Africa. This was James

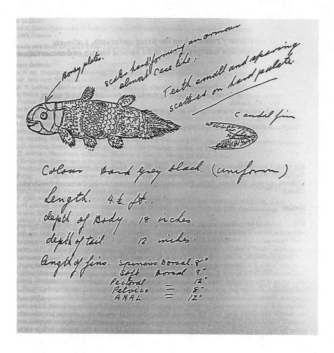

FIGURE 2   The sketch that Marjorie Courtenay-Latimer sent to J. L.
B. Smith. COURTESY OF J. L. B. SMITH INSTITUTE OF ICHTHYOLOGY

Leonard Brierly Smith, her correspondent and friend, 350
miles away in Grahamstown.

A keen zoologist and easily the most experienced active ichthyologist on the African continent, Dr. Smith lectured in
chemistry at the college. But his great love and passion was
fishes and the sea, starting from a youthful unfashionable interest in sea angling. Smith had devoted most of his life to
study of the fishes of southern Africa. All his spare time was
spent in expeditions along the coast up to Mozambique and
even the Red Sea. Like Latimer, he had earned the respect of
the trawlermen and commercial line fishermen by going out
and working on their boats in all weathers and all conditions.

Apart from Dr. K. H. Barnard at the museum in Cape Town
(who had completed a study of the fishes of South Africa in
1927 but then turned to the study of invertebrates), he was
really the only person whom Courtenay-Latimer could ask
about this new and wonderful fish.

In 1938 Smith was aged forty-one. His career had largely
been out of the public limelight. During service in the World
War I he had been a machine gunner in the Twelfth South
African Infantry. He contracted malaria, dysentery, and Malta
fever during the 1916 East Africa campaign and was in-
validated out, to return to his interrupted university studies at
Stellenbosch. His health was already in the precarious state
that dogged him for the rest of his life. He took a Ph.D. in
chemistry at Cambridge and returned to South Africa to teach
the subject at Rhodes University College in Grahamstown. His
second wife, Margaret, was instrumental in rekindling his old
love of fishes, but it was an uphill, difficult task to become a
serious ichthyologist while having to teach chemistry for a living. Nonetheless, starting with his first publication in 1931, he
had built up a good reputation in the field, principally through
a series of articles in the technical zoological journals in which
he described new fishes from the east coast of South Africa.
Many of his new species had been turned up by his own dili-
gent field researches, others by his broad range of friends and
acquaintances, and he knew that the field was essentially un-
tapped.

Smith was a lean and hard man—as spare in his habits as he
was in his body. He arranged his life to be functional, with no
luxuries or frills. His hair was cut in a military crew cut, and he
cared little about his clothes. He was very careful about his
diet, however, for many years alternating days when the only
meat he would eat was fish with days when he would eat only
fruit and nuts. He was moody and intense. He was probably a
difficult colleague, too, but at the same time respected for his
drive and dedication. The one thing he was passionate about
was fishes, and in this his second wife, Margaret, was his constant companion and fiercest publicist, working in the field

alongside him and illustrating many of his published works
with her drawings and watercolors. She was once asked what it
was like living with JLB and replied: "A wife can be independent or indispensable, but not both; I choose to be indispensable."[9] It was a good partnership, and Smith needed such a
partner because most of the time, in his mind at least, the rest
of the world was either wrong or contrary, or both.

Much of Smith's research and writing was done at a laboratory he had built at his family place at Knysna on the southern
Cape coast (now a very popular resort area). In late December
1938 the Smiths were there for the Christmas vacation. As was
increasingly the case those days, Smith felt unwell and he was
also weighed down by a huge number of examination papers
to be marked. In fact, he had obtained permission from the
college to stay on at the coast to get his grading completed.
Courtenay-Latimer sent her letter to Grahamstown; owing to
the holidays and to its being forwarded, the letter did not
reach Smith until January 3.

Really great discoveries require a combination of good fortune, learning, imagination. But they also require intellectual
courage and even physical stamina. They are a terrible responsibility and require a tough person to handle them. In many
ways Smith was temperamentally unsuited to the burdens of
major discovery. He was an eccentric, a loner, intense, brood-
ing, ridden with anxieties. In other ways, however, he fitted the
task. He was deeply immersed in South African ichthyology
and very broadly read in the study of all fishes. No stuffy academic, he was always prepared (perhaps even eager) to take an
unpopular or unconventional position if he was sure of his
ground. While he professed to hate publicity, he also courted
it and was a master at dealing with the press. The discovery of
Latimeria would consume, indeed nearly kill, him, but it also
gave him an opportunity to become a world figure in zoology.

Courtenay-Latimer's letter arrived, and as soon as Smith
looked at the sketch, he saw from the limbs, the tail, and the
scales that this was something new and very different—a fish
unknown to science. He had the same impression as Cour-

tenay-Latimer, that this was something like a lungfish. But the thought that it was something far more exciting quickly grew in his mind. He started to run through in his mind everything he had read or seen about primitive fishes. And then, as he wrote, "a bomb burst in my brain . . . and . . . I was looking at a series of fishy creatures that flashed up as on a screen, fishes no longer here. . . . I told myself sternly not to be a fool but there was something about that sketch. It was if my own common sense were waging a battle with my perception, and I kept staring at the sketch."[10]

As he wrote later, he was so transfixed by the letter and the sketch that his wife asked him in alarm what was the matter. "I said quite slowly, 'This is from Latimer, and unless I am quite off the rails, she has got something that is really startling. Don't think me mad, but I believe there is a good chance that it is a type of fish generally thought to have been extinct for many millions of years.' "

If only he was right, this was the zoological find of the century. Smith had not studied many fossil fishes in person; but he knew the literature, and he thought that he knew where there was a picture of a fossil fish that looked very much like this. And if it was the same, the fish could only be a *coelacanth,* technically a member of the *extinct* order Coelacanthini in the subclass Crossopterygii (as it was then called)—a lobe-finned fish related to fishes that once lived in the Devonian age (more than three hundred million years ago). The courage and imagination required to envisage this were immense, because it was only too true that the only other coelacanths that anyone had ever seen were fossil ones. In fact, the whole group was thought to have been extinct for at least seventy million years. The last coelacanths had died out with the last dinosaurs. Before Captain Goosen, the crew of the *Nerine,* and Courtenay-Latimer, no Westerner had ever seen a living coelacanth.

Smith was pretty sure of what it was. But there was too much to do. He needed to get the right book to check his memory. He needed to get back in touch with Courtenay-Latimer to make sure she had saved the fish. But he also needed to keep it a secret. If he were to say anything to anyone and then turn out to be wrong, he would be a laughingstock. If he were right, he did not want to share it with anyone. The pressures would be enormous, he knew, and he knew himself well enough to be afraid. "I was afraid of this thing, for I could see something of what it would mean if it were true, and I also realized only too well what it would mean if I said it was and it was not. On that sketch alone I could never decide anything; I must journey to see the creature itself. That would almost certainly mean a journey to East London." But he was physically tied to Knysna by the examination papers and emotionally tied up as well.

So he did not immediately dash off to East London, a journey that would in any case have been difficult. First he cabled back to Courtenay-Latimer, then followed up with a letter. And he wrote to Barnard at the South African Museum in Cape Town (a friend of his) for a copy of Volume II of Arthur Smith Woodward's *Catalogue of Fossil Fishes of the British Museum (Natural History).* [11] Here is an extract from his letter to Courtenay-Latimer, cautious with respect to an identification yet firm about the importance of saving the fish.

After admonishing her to save the gills and viscera, he writes: "I cannot hazard even a guess at the fish at present, but at the earliest opportunity I am coming to see it. From your drawing and description the fish resembles forms that have been extinct for many a long year, but I am very anxious to see it before committing myself. It would be very remarkable should it prove to be some connection with the prehistoric. Meanwhile guard it carefully, and don't risk sending it away. I feel it must be of great scientific value." [12]

Now began the period of intense personal agony for Smith:

> My worries carried me along, my mind was in a chaotic state. Was this a prehistoric relic? . . . My mind was busy all the time trying to assign that sketch to some clear type. It appeared to be something like a shark in its make-up, but so were those early Crossopterygians. . . . Those were awful days, and the nights

were even worse. I was tortured by doubts and fears. . . . Yes, everything was against its really being a coelacanth . . . and yet every time I took out that sketch, it said "yes," emphatically "Yes."[13]

It would have been easier if he had the fish in front of him, in a modern laboratory at the British Museum, with every reference book available and thousands of fossils in a great collection for comparison. But there he was, stuck in a remote seaside cottage, and he was, in any case, practically the only ichthyologist on the continent. On January 6 the Smith Woodward book arrived (Barnard was prompt). Smith found the descriptions of the fossil coelacanths, a lithograph of a fossil called *Macropoma* from the Cretaceous and a drawing of the Jurassic *Undina* that were for all the world like Courtenay-Latimer's sketch. That basically clinched it. Furthermore, nothing else would be available to help him out.

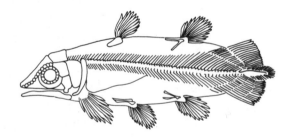

FIGURE 3   Smith Woodward's drawing of the Jurassic coelacanth *Undina* (simplified).

But for days and then weeks he could not bring himself to face the final decision of going to East London and seeing the specimen for himself. It would still be his judgment and his reputation at stake. He did test the waters by writing to Barnard again and telling him what he thought he might have. Just as he feared, Barnard was disbelieving. (The letters have apparently been lost.) He wrote again to Courtenay-Latimer on

January 9 and was really quite specific to her that "it is almost certainly a Crossopterygian allied with forms that flourished in the early Mesozoic or earlier."[14] But he stayed in Knysna, marking examinations and fretting.

Meanwhile, pressures had been building in East London. The Christmas season in East London is hot, and the fish was big. No container at hand was large enough for Center or Courtenay-Latimer to preserve it in formaldehyde in the usual way. On December 26 they checked the fish. The formaldehyde cloths were not working; indeed, it seems naive that they would have thought they would work. The fish was also oozing great quantities of oil. Courtenay-Latimer had heard nothing at all from Smith, but decisions had to be made. There was no choice, considering the resources of the little town. The taxidermist would have to make a trophy mount. This meant that the fish would be skinned, and while most of the bony skull would be preserved to make the shape of the head, the rest of the body, the muscles, and the decaying viscera would not be saved. The tongue apparatus was cut out, later to be cast in plaster and the cast installed in the open mouth of the mount. It was small enough for Courtenay-Latimer to take back to the museum and put it in formalin.

Finally, on January 3, at 10:00 A.M. Smith's first cable arrived: "Most important preserve skeleton and gills fish described."[15] But it was too late. All the "waste" body parts had gone into the trash, and even though they tried looking in the municipal dump, nothing was found (everything was probably disposed of in the ocean). Just as bad, it turned out that the photographs had not come out; the film had been spoiled. Moreover, none of Courtenay-Latimer's board of trustees seemed very interested in what was happening.

Smith and Courtenay-Latimer exchanged anguished letters over the loss of the soft parts. But at least Courtenay-Latimer could send some more details from the taxidermist's observations: "There was no skeleton. The backbone was a column of soft white gristle-like material, running from skull to tail—this was an inch across and filled with oil—which spouted out as cut through—the flesh was plastic, and could be worked like clay—the stomach was empty . . . the gills have small rows of spines . . . oil is still pouring out from the skin, which seems to have oil cells beneath each scale. The scales are armor like fitting into deep pockets. . . ."[16]

FIGURE 4   The living Australian lungfish *Neoceratodus.*

On January 17 and again on January 24 Smith wrote to Barnard in Cape Town stating that he was virtually certain he had a coelacanth. Barnard was still incredulous and said that he had mentioned it in confidence to Dr. E. D. Gill, the director of the museum. Gill had himself worked with fossil fishes, including an important paper on fossil lungfishes written with Professor D. M. S. Watson of London University. But apparently Gill was also of the opinion that Smith was dangerously deluded. After all, there was no evidence to show them, no photographs of the fish, for instance. By now Smith realized that he really was going to have to do this all by himself, and he was casting about wildly for rationalizations to justify his identification. For example, he tried to persuade himself that this was some sort of mummified specimen of an extinct species left over from the Mesozoic in the ooze at the bottom of the ocean (others were later to try the same argument), but no, that wouldn't work: It had snapped its mouth when Goosen touched it. Then, on February 1, Courtenay-Latimer sent him some scales. In fact, although they are tough and leathery, the scales of coelacanths dislodge from the skin very easily, and it is surprising that she had not sent any before. At last Smith had tangible evidence, and there no longer could be any doubt at all. This fish existed, it was real, and it was a coelacanth. The scales were exactly like those described (by Smith Woodward, for example) from the fossil coelacanths; no other kind of fish had scales like them.

On February 8, 1939, six weeks after the fish had been caught, Smith and his wife finally left Knysna for East London via Grahamstown, but with rains and bad roads, it was a week before they could complete the journey.

Courtenay-Latimer remembered: "It was 16 February 1939. We had been having terrific rains. . . . As he [Smith] walked into my small office where I had the now mounted fish he said, 'I always KNEW somewhere, or somehow a primitive fish of this nature would appear.' "[17] Smith's own account is a little different:

> We went straight to the Museum. Latimer was out for the moment, the caretaker ushered us into the inner room and there was the Coelacanth, yes, by God! Although I had come pre-

FIGURE 5   Fifty years later Miss Courtenay-Latimer photographed on the dock where she collected the first *Latimeria*. COURTESY OF DR. RICHARD GREENWELL AND THE INTERNATIONAL SOCIETY OF CRYPTOZOOLOGY

FIGURE 6 Smith's formal photograph of *Latimeria* has an arrow indicating the position of a structure called the spiracular organ, and the torn rear margin of the first dorsal fin has been shaded in.

pared, that first sight hit me like a white-hot blast and made me feel shaky and queer, my body tingled. I stood as if stricken to stone. . . . I forgot everything else and just looked and looked, and then almost fearfully went close up and touched and stroked, while my wife watched in silence. Latimer came in and greeted us warmly. It was only then that speech came back, the exact words I have forgotten, but it was to tell them that it was true, it really was true, it was unquestionably a coelacanth. Not even I could doubt any more.[18]

There is an odd footnote to be added here. The story that Courtenay-Latimer was not present when Smith arrived after this six weeks' delay was apparently due to Mrs. Smith, who even suggested that Courtenay-Latimer was out shopping. All this is hotly denied by Courtenay-Latimer, who was also to become upset when, on another occasion, Mrs. Smith stated that Courtenay-Latimer had allowed the taxidermist to dispose of the rotting viscera because she "assumed that the fish could not be important."[19]

The next day a reporter appeared. This was something else

that Smith feared. In his intensely personal preoccupation with the identity of the fish, which he finally had seen and was certain about, Smith now began to be consumed with worries about how he would release news of the discovery to the world. He needed to release enough information to be believed, but he didn't want anything to be written that would in any way allow him to be scooped. For example, it was important to reserve for himself the honor of giving the fish its formal scientific name, and he had already decided what that name would be: *Latimeria chalumnae,* to honor Courtenay-Latimer and to mark the place where it had been found. He planned to write a short notice to the prestigious scientific journal *Nature* in London, announcing discovery to the scientific world. And then he would prepare a full and lengthy monographic description, a task that would take months or years. But if news were to appear first in the papers, perhaps some unscrupulous rival would try to give the fish the new scientific name by which it would forever be known and thereby become the person with whom it would forever be associated in the scientific literature. This might be possible, using the sort of information a newspaper article might include. (Such opportunism was not unknown even in the civilized world of academics; in the forties there was one particularly notorious academic "pirate" in Australia who made a practice of combing the scientific literature for lapses that could be pounced upon for the opportunity to name a new species.) Smith needed time to work with the fish, to prepare these momentous papers, but the press could not wait. Too many people already knew something of the matter, in East London at least. So finally he agreed to a short article, but he also tried to insist that there be no photographs. Courtenay-Latimer knew there had to be photographs. The reporter, of course, knew that without a photograph no one would believe a story coming from East London, of all places. He pleaded for permission to take photographs and promised they would not be published anywhere except in the *East London Daily Dispatch,* to which Smith finally agreed.

The result was predictable: In the *Dispatch* of February 20 there appeared a rather sensational article complete with two photographs, and soon, complete with the same photographs, it was taken up by wire services all over the world. The reporter apparently made a fair business out of selling the photographs over the next few months, even having the gall to offer copies to the East London Museum for two pounds! Now the coelacanth was general news; both it and Smith were public property. Marjorie Courtenay-Latimer was already receding to the sidelines.

By the beginning of March Smith had already sent a preliminary scientific notice to *Nature,* complete with photographs, and from that point on there never was any further objection to Smith's identification of this fish as a coelacanth. This article was published on March 18, 1939, with the title "A Living Fish of Mesozoic Type."[20]

However, Smith was somewhat scooped by an article published a week before (on March 11) in the *London Illustrated News.* This article—"One of the Most Amazing Events in the Realm of Natural History in the Twentieth Century"—complete with photographs that had arrived at the British Museum at the beginning of the month, was written by Dr. E. I. White, paleontologist at the British Museum (Natural History) in London.[21] On March 16 J. R. Norman, ichthyologist at the British Museum (Natural History), read a paper on the discovery of the coelacanth at a meeting of the Linnean Society in London.[22] This talk was an exposition of the paper that Smith had sent to *Nature.* After Norman had ended his presentation, Sir Arthur Smith Woodward (director of the museum and the author of the definitive catalog of fossil fishes that Smith had consulted) commented on the similarity of the living coelacanth to fossil forms such as *Macropoma.* He commented that extreme conservatism of structure between Cretaceous forms and their modern descendants was observable in other groups as well, such as the fish families Berycidae and Halosauridae. Professor D. M. S. Watson, from University College London, stressed the value of the discovery as a test of the accuracy with which paleontologists were able to reconstruct and interpret fossils, and Dr. White added that it was possibly the most important zoological discovery since the Australian lungfish, which had been discovered (and thought at first to be an amphibian) in 1870.

This meeting of the Linnean Society rated almost a full column coverage in the *Times* of London of March 17, and now the world's press took up the story.[23] Scientists everywhere debated the case, but generally Smith found that only in South Africa was he doubted. Overseas the news was accepted as true, and no doubt the photographs were crucial to this. Ichthyologists all over the world could see the evidence although they had to believe that it was not a fake made up out of Plasticine and wire. Smith's reputation as a good field ichthyologist prevented that.

Meanwhile, the East London Museum trustees had suddenly become very interested in *their* famous fish. A special viewing day was arranged for the public. It is reported by Courtenay-Latimer that more than twenty thousand people visited the museum (this figure is surely exaggerated). In light of this interest, it is perhaps remarkable that Smith was able to persuade the very reluctant trustees of the museum to have the specimen shipped by rail to Grahamstown, where he could work steadily on preparing proper scientific descriptions. The fish arrived under police escort! Now the publicity value of the fish was being exploited by all parties.

Smith had no chance of getting relief from his duties in the chemistry department, so he worked at his projected great monograph on the coelacanth every day from 3:00 A.M. to 6:00 A.M. Then he would go for a four-mile "walk in the hills," come back, write up his observations, and leave for the college at 8:30 A.M. In the evening he worked again until 10:30. All the intensity and single-mindedness of Smith's character came to the fore. "It was the same old mixture my life had always been, turbulence and trouble, only more intense. We had no social life, business and financial affairs took a back seat, and our food reached its destination over and between sheets of manuscript."[24] The fish, which Smith characteristically kept in the house in a separate room rather than at the university, totally dominated the Smiths' lives, even with its peculiar smell, just as Smith had known it would.

Smith was quickly inundated with the usual crank mail, with which he could deal fairly readily. His scientific colleagues around the world were a different matter. Inevitably other scientists would start to write articles based on his descriptions and would vie to interpret what the discovery might mean. It was equally inevitable that Smith would react with hostility. The first to call forth his wrath was White's article in the *London Illustrated News*. "I did not find it flattering to remote scientists like myself,"[25] said Smith. White could be extremely charming, but he was also blunt-spoken and often tactless, perhaps in part because of the pain he suffered from a back injury. He was usually very sure of his own opinions. He wrote in his confident way that "the cause of survival of an archaic type is undoubtedly competition with more highly developed and efficient forms. Under pressure of such competition, the older forms are forced to retreat to a less and less advantageous environment. . . . Our living coelacanth . . . almost certainly was a wanderer from the deeper parts of the sea. . . ."[26] White's use of the word *our* may have rankled as much as his categorization of the coelacanth as a less efficient form. But Smith mostly disagreed with the notion that it came from the deeper sea, as we shall see below.

Smith sent off a second report to *Nature,* published on May 6, 1939, in which he provided more details about the fish, including a photograph of a microscopic thin-section of a scale. He was able to report that, like the lungfishes, the fish had an air bladder or lung (on the basis of the taxidermist's report of the discarded viscera), which was a median rather than paired structure. He described many of the important features of the skeleton and some of the sensory organs. Important among the latter was a curious set of structures in the snout. In addition to more or less normal nasal organs, there was a large median chamber, with three pairs of tubes opening to the surface. This organ, which we now call the rostral organ, was thought by Smith to be part of the nasal apparatus, but we now know it belongs to a separate sensory system (see Chapter 8). Smith also used this report to answer criticisms of Courtenay-Latimer concerning the loss of the viscera: "Few persons outside South Africa have any knowledge of our conditions . . . it was the energy and determination of Latimer which saved so much and scientific workers have good cause to be grateful. The genus *Latimeria* stands as my tribute."[27]

An interesting footnote to the early flurry of opinion concerning *Latimeria* was that J. R. Norman, in his paper to the Linnean Society of London of March 11, quite reasonably suggested that the fish was an inhabitant *either* of much greater depths than where it had been caught *or* of places where the sea bottom was rocky and trawling impossible. Prophetically he suggested that "long-baited lines perhaps hold out the best chance of catching fresh specimens."[28]

In order to make the fullest possible scientific description, Smith needed to salvage every possible scrap of data from the poor mount that Center had made. In his laboratory Smith opened up the mount and dissected it from the reverse side. To his delight, most of the delicate bones of the head were still there under the skin. This was tedious work, and without relief from his teaching duties, it would go slowly. Even if he were to have had full time to devote to his researches, Smith would hardly have wanted to work very quickly; his approach was far too painstaking.

Within a couple of months, however, the trustees of the East London Museum were agitating to get the specimen back. From their point of view, Smith seemed to have annexed the fish for himself; no one else could even see it. (The trustees would have been horrified to see it in its dissected state, no doubt.) The trustees had also began to discuss the possibility of selling the specimen. Perhaps they were tempted in part by enterprising types who could see the gains to be made in exhibiting the specimen around the world as a sort of sideshow freak. There were well-endowed museums in Europe and

North America that would have given anything to have the specimen in their collections. Furthermore, influential voices at home argued that a specimen as important as this should be at some central and important place like the British Museum. At the very least it should be in the South African Museum in Cape Town. Certainly it should not be hidden away in Cape Province.

This was the sort of opinion guaranteed to enrage Smith, of course. Apart from that, he needed much more time for his studies. But the trustees would agree that he could keep it only until May 2. So he worked on in "a frantic nightmare."[29]

The specimen was returned to the East London Museum in May, and later it was completely remounted by the best taxidermist available, Mr. Drury from the South African Museum. Placed on display in East London, it attracted great crowds. Then, in June, the chairman of the East London trustees gave Courtenay-Latimer a letter to type. It was a letter to the British Museum (Natural History) offering it the specimen. She read the letter and decided that she could not possibly type it, that she would resign first. Several days later, when Dr. Bruce-Bays came to ask her for the letter, she exploded, telling him "in no uncertain terms what she thought of the whole thing."[30] Bruce-Bays was overwhelmed by the force of her feeling and her arguments, and from that moment the matter was never heard about again. So the specimen stayed in East London where it can be seen today. All this is just as well because had it gone to Britain, over the following years there would surely have been enormous pressure to get it back and much bad feeling as a result.

Smith's monograph, "A Living Coelacanthid Fish from South Africa," was published in February 1940 in the *Transactions of the Royal Society of South Africa.*[31] It laid out a remarkably complete description of *Latimeria chalumnae,* considering the state of the specimen. Smith was able to describe the skeleton very fully. Among important features he revealed was that the skull was of the same structure as fossil coelacanths and many of the extinct crossopterygian lobe-finned fishes in possessing a curious internal joint, the intracranial joint, that divided the braincase into two halves, front and back. Associated with the floor of the braincase in the region of this joint was a muscle sheath, the function of which was unclear. There was no sign of an internal nasal opening, or choana, of the sort that tetrapods or lungfishes have. The skull totally lacks the marginal maxillary bones that most vertebrates have, and the only teeth are those on the palate and the lower jaws. Smith noted that the inner surfaces of the scales show growth rings, from which he estimated that the fish was twenty-two to twenty-five years old. Because of these growth markings, the fish was unlikely to have lived in the great seasonless depths of the ocean. An interesting feature of this monograph, with its 106 pages of text and 44 photographic plates, is that unlike most scientific treatises, it has no bibliography. Throughout, Smith makes not a single reference to the work of anyone else.

Sir Arthur Smith Woodward published an extremely laudatory review of Smith's monograph in *Nature* (July 13, 1940) in which, however, while going out of his way to compliment Smith on the way in which he managed to recover so much information from the specimen, he makes the curious (and basically gratuitous) remark that "when the specimen was sent to the East London Museum its scientific value was not appreciated, and it was entrusted to a taxidermist. . . ."[32] It seems to have been impossible for Smith and Courtenay-Latimer (and, indeed, Captain Goosen) to get a straightforward appreciation of what they had accomplished under their circumstances. These criticisms of the preservation of the specimen were still causing ill feelings fifty years later.[33]

In the following chapters we will put the living coelacanth into its proper scientific context, but we should note that coelacanths were of great interest to zoologists not only because they represented an extinct form of fish or because the whole group of fishes to which they belong is very primitive but, perhaps most interesting of all, because this group of lobe-finned fishes that Smith and his contemporaries called the Crossopterygii (fringe fin) had in Devonian times included the immediate ancestor of all the land-dwelling vertebrates. Paleontologists traced the origins of the amphibians, reptiles, birds, and mammals back to that group. No one knew exactly what the ancestor was (in fact, we still do not know). But it was reasonably certain that while the coelacanth group was not itself the ancestor, it was closely related to the ancestor. To find a living example of any group thought to have been extinct for seventy million years was unusual, to say the least. To find a living crossopterygian fish was extraordinary, and as we will see in the next chapter, it held the possibility of illuminating all sorts of questions about the early evolution of fishes and the origin of land vertebrates including, eventually, man.

With the publication of his monograph, Smith was established as a significant figure in the world of zoology even though he considered that his work on the coelacanth had just started. Having discovered that the fish existed was only the beginning. Where there was one, there must be more. But where? He did not believe for a moment that the fish came from the deepest abysses of the ocean. With all his experience of marine fishes, it looked to him like a fish that lived on the deeper parts of the reef, but probably in water no deeper than a few hundred feet. For example, it was entirely the wrong color for a fish living deeper. But in that case why had specimens not been caught before? Significantly, while there had been the usual cranks, none of his reliable friends among the sea anglers and none of the commercial fishermen who trawled the South African coasts, most of whom he had worked with over the years, had gotten in touch with him to say, "Oh, yes, we catch that fish now and again." There was one man who swore that he had seen a similar fish washed up on the beach years before, but there was no record that could be authenticated.[34] This must mean that it did not live in the more shallow waters where most fishermen worked locally and that probably it was a stray to the South Africa coast. To be sure of this, Smith had photographs made and distributed to all possible fishing operations. The South African Fisheries Commission sent to the area around East London a research vessel that

fished intensively using trawls to try to catch another specimen. But these ventures turned up nothing, and Smith became
more and more confident that the fish was not native to South
African waters.

If *Latimeria* were not a truly deep-sea fish, strayed from the
depths of the Mozambique Channel between South Africa and
Madagascar, then it must have come from the coasts farther
north or from the islands scattered in the western Indian
Ocean. To find the source population(s) of this fish would be
an immense task. Most of the African coast was still scientifically unknown territory. A number of the inshore fishes were
known, together with a good selection of those fishes that were
commonly taken by native fishermen for food. But nothing was
known of the deeper coastal waters except for those areas
where South African trawlers worked. The islands in the western Indian Ocean were an even worse problem. To the immediate east was the great island of Madagascar, where the
French had a long-established colonial regime but the fish
faunas of which were still not well known. West and north of
the tip of Madagascar was a scattering of islands, reefs, and
banks from the Comores to Aldabra, and farther distant a
north-south chain from the Seychelles to Mauritius and Réunion, linked by the relatively shallow underwater Mascarene
Ridge. Even farther to the east was another line of islands
running almost due north-south—Laccadives, Maldives,
Chagos Archipelago, reaching up to the Indian subcontinent.
Farther south there were scattered islands in the colder waters
of the southern ocean. The fact that the prevailing currents
run south down the Mozambique Channel suggested to Smith
that he must look north and east.

Marjorie Courtenay-Latimer took the East London coelacanth down to Cape Town to be remounted at the South
African Museum in September 1939. Coming home, she got
off the train to learn that war had been declared with Germany.
Very soon all serious work in ichthyology had to stop because
of the war effort. For the next five years Smith's work in chemistry had to come first. The university had doubled up its

classes, leaving little time for fishes, although Smith continued to devote whatever time he and his wife could find to collecting and studying their beloved fishes from the South African coasts.

With the end of the war Smith's scientific fortunes improved. Like most developed countries, South Africa had learned from the war that there was a tremendous need to invest in science and technology, especially by training new generations of young people in basic science and encouraging all sorts of research initiatives. Science was less and less an amateur's game where great discoveries could be made merely with the aid of sticks, sealing wax, and string or where a great scientist needed only a supply of yellow notepads and pencils. The South African government established a Council for Scientific and Industrial Research (in parallel the United States established the National Science Foundation, and Great Britain the Department of Scientific and Industrial Research). Research became a national priority, and science became a matter of national prestige. Smith received one of the first grants, a fellowship that would release him from a need to earn his living by teaching chemistry. The university set up a department of ichthyology, largely funded from the council, and at last Smith was free to devote himself to ichthyology full-time. Moreover, he had been asked to write a book on the fishes of South Africa, a project he had started once before but had lacked time and funding to complete. Now his work on South African ichthyology proceeded apace.

Internationally interest in the coelacanth picked up steadily after the war, and in South Africa a committee was formed for the purpose of mounting a major oceanographic expedition both to search for coelacanths and generally to explore the oceanography of the Mozambique Channel. As such committees often do, it soon got bogged down, and in particular there developed a conflict between Smith, who wanted only to look for coelacanths and knew exactly what he wanted to do, and the rest of the committee, who wanted to use the expedition to further their own pet projects. In any case there was a no suitable ship in South Africa, and when Prime Minister Jan Smuts was sounded out on the possibility of a cooperative effort with the British research vessel *William Scoresby,* which was headed for South Africa to fish for coelacanths in 1946, he vetoed the idea. To Smith's disgust, without his participation the *William Scoresby* expedition produced nothing, and Smith's instinctive distrust of "government" and politicians was reinforced.[35]

However, one positive step was taken. Smith got the committee to agree to have a poster prepared (in Portuguese, English, and French), to be distributed all over the western Indian Ocean, from the French territories of Madagascar and the Comoro Islands, to Portuguese Mozambique and British East Africa, as far north as the Red Sea. It carried a photograph of the coelacanth and offered a reward of one hundred South African pounds to anyone who could catch one. This poster was sent to government officials and fishing concerns all over the region although Smith and others wondered whether regional government officers would really take the trouble to distribute it.

Meanwhile, other organizations tried expeditions to the region, with no success at all. There was a fruitless Danish expedition and a modest South African experimental fishing program. Smith was supremely frustrated being sure what he would do with greater resources. Lacking them, he had to plug away on his own terms, and for Smith that meant grass-roots fieldwork, getting into the field at every opportunity, traveling to remote coasts, talking to natives, to fishermen, all the time collecting more fishes. All this backbreaking work was important for his big *Sea Fishes* project, and sooner or later surely another coelacanth would turn up.

In 1949 Smith's *Sea Fishes of Southern Africa* was published, and now, fully established in his international scientific reputation and with his position at home finally secure, Smith became more and more preoccupied with the coelacanth question.[36] Simply, where were the coelacanths? They had not turned up in several expeditions to Mozambique, so Smith and his wife went farther north.

FIGURE 7   Smith's poster offering a reward in Portuguese, English, and French. COURTESY OF J. L. B. SMITH INSTITUTE OF ICHTHYOLOGY

In 1952 Smith mounted a big expedition to Zanzibar Island, Pemba Island, parts of Tanganyika, and Kenya. Everywhere the Smiths went they not only collected fishes but talked about the search for the coelacanth to reporters, local officials, and anyone who would listen. Smith was a wonderfully eccentric character for any reporter to write about, and the press followed the couple eagerly wherever they went. The Smiths held public demonstrations of their work. And everywhere they distributed the coelacanth poster.

FIGURE 8   Hunt and his crew on board *N'duwaro* in Dzaoudzi Harbor, December 1952. COURTESY OF J. L. B. SMITH INSTITUTE OF ICHTHYOLOGY

On Zanzibar they met a young man named Captain Eric Hunt, who operated the small trading and fishing schooner *N'duwaro* between the Comore Archipelago (then French territory) and the mainland. The Comores were one of the localities that the Smiths had long thought it would be important to check out, but they had not yet done so, concentrating on working north along the African main coast instead. Hunt was a tallish man in his thirties with a small mustache and an engaging grin. He was immediately captivated by the Smiths and fascinated by the search for the coelacanth. He studied the poster carefully. He was familiar with most ocean fishes from the region and knew the native markets. But he was sure that he had never seen a coelacanth. He had also not seen the poster before. Since he often went to the Comoro Islands, he promised to take some of the posters for distribution there.

In fact, there was a French research laboratory at Tananarive, on the west coast of Madagascar, with a resident fisheries staff, and it had already been sent the posters. According to Smith's later account, when Hunt was in the Comores in September, he found that, as Smith had feared, the posters seemed not to have been widely distributed. Perhaps they had been dismissed as a futile idea by the official who received them. When Hunt talked to the governor of the islands, however, he turned out to be extremely interested and had the posters distributed all around the Comoro Islands immediately.

Upon completing their East African expedition in early December 1952, Smith and his wife embarked at Mombasa on the Union-Castle steamship *Dunnotar Castle* to return with all their collections and field gear to Cape Town. They called in at Zanzibar again on December 14, and Mrs. Smith went ashore to visit the local market. Hunt's schooner was anchored in the harbor. He had just returned from his trip to the Comores and was soon heading back in that direction again. He told Margaret Smith that he had distributed the posters and half-jokingly asked her what he should do if he actually got a coelacanth because there was no refrigeration in the Comores and probably no formalin. She told him that in that case the only thing to do would be to try to preserve it in salt. As they parted, he joked: "OK thanks, Anyway, when I get a coelacanth, I'll send you a cable."[37]

CHAPTER

2

To the Comores

"une très belle pièce. . . ."
—*Affane Mohamed, quoting
Abdallah Houmadi*

Although Professor and Mrs. Smith did not know it as they sailed south, the process of history had begun to repeat itself.

On December 24 the *Dunnotar Castle* reached Durban, and among the many letters and telegrams waiting for Smith was one urgently pressed into his hand by a ship's officer. It was from Hunt and had been forwarded from Grahamstown. It evidently was a follow-up to one that had been sent earlier. It read: "Repeat Cable just received have five foot specimen coelacanth injected formalin here killed 20th advise reply Hunt Dzaoudzi."[38]

A coelacanth! Excitement and even panic ensued. "My

heart turned right around or it felt like that . . ." Smith wrote
later. After fourteen years of waiting, it seemed impossible.
Smith did not even know where Dzaoudzi was, but one of the
ship's officers quickly discovered that it was part of a small
island called Pamanzi, just off the bigger island of Mayotte in
the Comore Archipelago. It was the site of the French colonial
administration for the islands and had an airstrip. Smith fired
off a reply: "If possible get to nearest refrigeration in any case
inject as much formalin possible cable confirmation that specimen safe Smith."[39]

Once again Smith was under intense pressure. The situation could hardly have been worse. He needed to get to the
Comores immediately, and that could be only by plane. There
were no commercial flights. How could he get access to a
plane? He was far from home, and all his field gear was deep in
the hold of the *Dunnotar Castle.* It was Christmas Eve. And the
fish had been dead four days already! Had Hunt really found a
coelacanth? And did he really have access to formalin to preserve it? "For a while I went almost insane," wrote Smith.

Soon a copy arrived of the original telegram that had been
sent to Smith's office at the university: "Have specimen coelacanth five feet treated formalin stop absence Smith advise or
send plane immediately—Hunt Dzaoudzi Comores."[40]

Hunt was right; they needed a plane. And their only hope
seemed to be the help of the government. One by one they
went through their possibilities and kept coming back to a
single chance; to reach the prime minister, Dr. Daniel Malan,
directly. But Smith remembered only too bitterly his rejection
by Malan's predecessor, Smuts; once again he hesitated at the
crucial moment. Lacking a clear course of action that he was
willing to take and racked with doubt, Smith was paralyzed.
Friends advised and tried to help, but no plan seemed to work.
Christmas Day came and went, and on Boxing Day (December
26) a new and more ominous cable arrived from Hunt: "Char-
ter plane immediately authorities trying to claim specimen but
willing to let you have it if in person stop paid fisherman reward to strengthen position stop inspected [presumably this is

a typographical error for injected] five kilo formalin no refrigeration stop specimen different yours no front dorsal or tail remnant but definite identification Hunt."[41]

This cable produced new agonies. With all the excitement at that end, naturally the French authorities would want to keep the fish, but what did Hunt mean about the fins? It sounded as though this were a second and new kind of coelacanth. But perhaps Hunt was wrong and it was not really a coelacanth. After all, what did he really know about fishes? Suppose Smith managed to get a plane from the air force and then the fish turned out to be merely a misshapen rock cod. It was the agony of January 1939 all over again.

Finally Smith consented to an attempt to get help from the president of South Africa directly, and contact was made with Malan's office. Malan was on vacation, but at about 11:00 P.M. on December 26 he called Smith back and, to Smith's surprise and relief, at once agreed to lend an air force plane. Smith's reputation as one of South Africa's leading scientists had carried the day. And by the strangest chance, one of the books that Mrs. Malan had taken for reading on vacation at the seashore was Smith's *Sea Fishes,* a complimentary copy of which had recently been sent at Smith's request. With that, all hesitation fell away, and at 7:10 A.M. on December 28 Smith and a crew of six took off from Durban in a Dakota (DC-3) with destination the Comores.

This time there was little secrecy about what Smith was up to. The world's press was watching. For example, the *New York Times* for December 27 had a story headed "Prehistoric Fish Believed Caught," and on the thirtieth a story, "Air Race to Save Dead Fish Stirs Scientists Here," complete with interviews from zoologists and paleontologists at the American Museum of Natural History.[42]

In the version of the story published by Smith in *Old Fourlegs,* the second specimen of a living coelacanth was caught by a fisherman named Ahmed Hussein (his full name was actually Ahmed Hussein Bourou) and a fisherman friend whose name is recorded only as Soha, from the small village of Domoni on

the eastern shore of Anjouan Island. They had been fishing at
night from a dugout canoe in relatively shallow water (twenty
meters) and about two hundred meters from shore. When they
brought the fish up, Bourou smashed it on the head to kill it
and headed for the shore. The next morning he took it to the
local market, where luckily it was spotted by the head of the
primary school at Domoni, Affane Mohamed, before it had
been cut up or cleaned. The schoolteacher had seen one of the
famous posters and recognized what the fish was. He knew that
the instructions were to do nothing to the fish but to take it at
once to "some responsible person." That person turned out
to be Hunt, who had his schooner anchored at the harbor of
Mutsamudu on the other side of the island. Remarkably (or
perhaps not remarkably since the reward offered was a king's
ransom to these poor islanders) Bourou, Affane Mohamed,
and a group of Comorans were said to have carried the fish
that day twenty-five miles overland, across the mountainous
and densely vegetated island to Hunt. In later retellings this
trip to Mutsamudu has taken on epic qualities—"the path to
Mutsamudu led through 25 miles of deep valleys, where the
bush was dense and the way just a track, and over high mountains. The continual glaring sun and the humidity"—and so
on.[43]

Hunt had no access to formalin at Mutsamudu, and the local
doctor was away, so he did what Mrs. Smith had told him to do
and salted the fish, which was cut in half lengthwise in the
process. This meant that a fair amount of damage was done to
the head and internal organs, but most of it was saved. Hunt
then set sail for Pamanzi, where the regional government was
housed and there the French medical officer, Dr. le Coteur,
generously donated his entire formalin supply to inject all over
the specimen. Hunt cabled to South Africa and, while waiting
for Smith's reply and eventual arrival, had some tricky diplo-
macy to practice. His own livelihood depended on good relations with the French administration, but his loyalty was obviously to Smith. One can imagine the flurry of excitement that
the arrival of the fish caused among the French staff, livening

FIGURE 9   Captain Eric Hunt, photographed in the Comores next to one of Smith's posters, December 1952. COURTESY OF J. L. B. SMITH INSTITUTE OF ICHTHYOLOGY

up the otherwise dull routine of this distant colonial station. The governor, Pierre Coudert, immediately took a personal interest in the affair of this peculiar fish that was so important to Hunt and Smith and worth so much money for a reward. He was willing for Hunt and Smith to have it. After all, it was only a fish, and Smith had been the one to send out the posters. On the other hand, it might also be important enough for him not to allow it to leave the islands without doing some checking. He cabled the Research Institute at Tananarive, Madagascar; but it was Christmas Day, and the cable either went astray or was garbled. No reply came. Coudert came to the Solomonic compromise that if Smith came to get the fish, he could have it. But if he didn't come, then the government would keep it. How long, Hunt worried, would this agreement hold? Would Tananarive intervene?

Details of the story just given vary. A slightly different version with more details, is provided in an unpublished affidavit written by the schoolmaster, Affane Mohamed, in 1965.[44] This document, which was circulated by Dr. Jacques Millot (who had been at the Tananarive research station in 1952 and later

headed the French research effort on coelacanths), is a direct
response to Smith's book *Old Fourlegs* (first published in 1956).
Affane Mohamed states that he knew the poster about the coelacanth quite well because one has been posted on the wall of
his schoolhouse at the *beginning* of 1938 (*"vers le debut de l'année
1952, le Professeur Smith avid fait afficher dans tous les bureaux et
batiments administratifs . . . des croquis du Poisson fossile. . . ."*) So
unless Affane Mohammed's memory was faulty, the posters
that Smith had sent out *had* been circulated by the Comoran
authorities after all. In any case, Affane Mohamed had definitely seen a poster, and it made him interested in the case. He
started looking carefully at all the fish that were caught by the
villagers.

The day that the fish was brought in by Ahmed Hussein
Bourou was a big day because the islands were preparing for
les compétitions sportives de l'Archipel, to be held on the island of
Mayotte. Affane Mohamed was the captain of the Anjouan soccer team, which later that day would be traveling by boat from
Mutsamudu to Mayotte. That morning he went to get a shave
from the local hairdresser.

In their overnight fishing Bourou and Soha had caught
many fish, and they managed to sell all of them at the morning
market on the beach except the one that turned out to be a
coelacanth. There is an oddity here in that Smith later reported that the fish was well known to the Comoran fishermen,
who never kept them when they were caught because they had
no value for food; the local name was *gombessa* or *ngombessa.*
Affane Mohamed is quite clear that the fishermen did not recognize the fish either as something seen before or as a fish
depicted on the schoolhouse wall. In fact, it was the local hairdresser, Abdallah Houmadi, who spotted the fish at the market
and, although he also did not recognize it, told Affane
Mohamed about it: *"une très belle pièce qui ressemblait un peu à un
poisson huileux que les comoriens appellent Ngnessa mais dont les écailles
et les nageoires avaient une forme et un aspect bizarres."* (The "oily
fish" *ngnessa* or *ngessa* is the oilfish *Ruvettus pretiosus.)* Without
further ado, Mohamed set off for Bourou's house, where he

found the fish and instantly recognized it from the posters as a coelacanth. Even when he showed the fisherman the poster, Affane Mohamed had difficulty in persuading Bourou that it was important or could be worth a very great deal of money. But he got him to agree that it should be taken to the authorities at Mutsamudu.

In fact, no dramatic overland trek was necessary; arrangements had already been made for transport that day to Mayotte in connection with the soccer competition. So Bourou and his fish were put with everyone else onto a public works department truck that would take the team over the island to Mutsamudu, where the boat was waiting. The truck reached Mutsamudu at about 8:30 A.M. Ahmed Bourou was by now exhausted. *"Harassé de fatigue [il] jetait des regards deséspères sur son poisson, abandonne depuis une demi-journée à côté de lui."* In the end he did not continue with the others and was left to find his own way home on foot while the fish went on to Mayotte (so poor Bourou was the one who had to make a twenty-five mile trek).

The boat was, of course, Hunt's schooner the *N'duwaro,* as he was the trader who provided transport among the islands and between the islands and the mainland. So there is no mystery about how the villagers of Domoni knew there would be a "responsible person" over at Mutsamudu. Hunt quickly heard the news of the capture of this strange fish through *radio cocotier* (palm radio, or bush telegraph) and contacted Mohamed at midday. He immediately saw what the fish was and stepped in to take responsibility for it, saying that he would indeed take it to the authorities at Pamanzi. Mohamed agreed to this plan because no one else on Anjouan would agree with his identification and because Hunt took charge of trying to preserve it. They cut it open and salted it, then wrapped it in cotton and packed it in a crate. "Ownership" was subtly passing from Ahmed Hussein Bourou to Affane Mohamed to Hunt.

They left at 10:30 P.M. It was a lovely night—*"nuit serène avec un clair de lune très agréable,"* the soccer players singing songs as they sailed. Hunt invited Mohamed to join him in his cabin (very flattering to the Comoran schoolmaster), where they conferred about the fish. Hunt showed him Smith's paper on the first specimen, and even though the fins on this fish were different, there was no doubt that it was some sort of coelacanth. Interestingly, Mohamed states that (trans.) "in order not to divulge the story of the coelacanth too soon, Hunt begged me not to ask the administrator for compensation for the fisherman. On our return to Anjouan he would turn over the whole 50,000 francs promised to the interested party." Affane Mohamed was more inclined to let the administration have it. But (trans.) "unfortunately my interventions with the authorities were in vain, the coelacanth left for South Africa." The next morning *N'duwaro* arrived at Dzaoudzi; Hunt sent off his first telegram and began to formalize the fish with the help of the doctor. With that, control of the fish passed completely out of Affane Mohamed's hands.

Smith took off from South Africa on December 28 and reached the Comores just after dawn next day, after stops at Lourenço Marques and Lumbo, Mozambique. He was met at the tiny airstrip at Dzaoudzi by Hunt, whose first words, according to Smith's later account, were: "Don't worry. It's a coelacanth all right." A number of officials (and Affane Mohamed) were also on hand and Smith and Hunt were taken to the governor's residence, where Monsieur and Madame Coudert waited with a lavish reception. So far things were going well, except that poor Smith hated rich food and drinks (especially for breakfast) and was desperate to see the fish. He begged to go to the harbor, so they all walked down to Hunt's schooner, where the fish, wrapped in cotton in a crate, was waiting.

Once again came the moment of test for Smith. Would it be the real thing or some ghastly mistake?

> My whole life welled up in a terrible flood of fear and agony, and I could not speak or move. They all stood staring at me, but I could not bring myself to touch it [the crate]: and, after standing as if stricken, motioned them to open it, when Hunt and a

sailor jumped as if electrified and peeled away that enveloping
white shroud.

God yes it was true. . . . It was a coelacanth all right. I knelt
down on the deck so as to get a closer view, and as I caressed
that fish I found tears splashing on my hands and realized that I
was weeping and was quite without shame. Fourteen of the best
years of my life had gone in this search and it was true. It had
come at last.[45]

Now time was pressing. With bad weather approaching, the
pilot was worried about the journey back, but the governor's
reception could not be put off. They had to be polite, as Hunt
knew only too well. As they stood together on the schooner,
Smith said that this specimen seemed indeed to be different
from the first fish. It might possibly be a new genus and species, and he proposed to name it *Malania hunti.* Hunt was quite
alarmed at this and said that Smith must give it a French name;
so they agreed upon *Malania anjouae.* The fish in its crate was

FIGURE 10   This formal picture shows Smith, Hunt, Governor Coudert (white uniform), and the South African Air Force crew with the
second coelacanth. COURTESY OF J. L. B. SMITH INSTITUTE OF ICH-
THYOLOGY

put on a truck and taken straight to the airstrip, and Smith
walked back to the governor's residence with the others to
endure a reception of every delicacy that could possibly be
thought of to upset his delicate stomach—from wines to "an
enormous cake spread with sticky chocolate icing, the mere
sight of which made my liver throb. . . ." They talked of further
work in the Comores, about which the governor was very encouraging. Finally Smith made a little speech in which he
thanked the governor for his assistance and for letting him
have the fish, stressing that he really thought it was his fish
anyway because it had been his search. He said that he had
authorized Hunt to act as his agent and now offered a reward
of another hundred pounds for a third specimen, but if one
were caught, it would be given to the governor as the representative of the French nation. All seemed to be agreeable,
and after firing off triumphant cables to Malan and the Council
for Scientific and Industrial Research, Smith and the air force
crew took off again.

Naturally enough, Affane Mohamed's account of the same
events is a little less flattering to Hunt and Smith and more
than a little bitter. Evidently it was not until a plane actually
appeared in the sky that the dubious locals were persuaded
that this fish was important *"tous les sceptiques voulurent admirer la
prise car maintenant l'histoire leur paraissait sérieuse."* Mohamed
says that (trans.) "immediately after getting off the plane, Professor Smith went to the boat." Still wearing his flying gear,
Smith knelt, touched, and "observed our prize with profound
joy." Mohamed could not follow the conversation, which was
in English, and no one asked him any questions. *"J'assistais à la
scène en simple spectateur."* Then; "Our visitors were in a hurry to
leave, taking only a few minutes with the Chef du Territoire for
an apéritif." True to his word, Hunt returned on December 30
to Anjouan and gave the fishermen their reward of fifty thousand francs.

Smith was reasonably confident that other coelacanths
would now be caught off the Comores because, as noted
above, Hunt had discovered from the fishermen that the fish

was known to them (or at least to some of them). They knew its rough, bony scales quite well, and it was from Hunt that the story came that the Comoro islanders used the scales as sand-paper to roughen up the inner tubes of bicycle tires when they had to glue on a patch. (It is a sign of how infrequently people in the West repair bicycle tire punctures nowadays that some very recent accounts of the coelacanth have contained the information that the Comorans used the tough, prickly scales as the patches!) The fishermen perhaps caught one or two *gombessa* each year and usually at the sort of depth that Ahmed Hussein Bourou and Soha had been fishing—less than two hundred meters.

In his book Smith describes the trials of the flight home in the noisy, unheated, unpressurized DC-3, lying on the floor of the plane next to the precious crate, enduring yet another night without sleep. While everything was agony for Smith, evidently the plane's crew members thoroughly enjoyed their offbeat mission, tensions and all, despite the fact that they all had been taken from their families during the Christmas holi-day. They respected this half-crazy, obsessed ichthyologist who ate only fruit and nuts and whose idea of a good time was apparently to travel to some godforsaken place, live in a tent on the shore, and collect smelly fishes. Typically they flattered him with a practical joke. They passed him an urgent note: "Managed to intercept a message stating that a squadron of French fighter planes left Diego Suarez [site of a major French military base on Madagascar] before we left Dzaoudzi with or-ders to intercept us and compel us to return to Madagascar." "What speed can they do?" Smith anxiously shot back," . . . any hope of escaping in a cloud? Well . . . I don't know how you chaps feel about this, but I'm not going back. I don't believe they would dare to shoot us down. . . ." Then the crew burst out laughing but it was some minutes before Smith realized that it was a joke.[46]

They got the fish back to South Africa safely; the prime minister was pleased (when he actually saw the fish, he said, "It's ugly"); Smith was ecstatic. The world was once again enthralled by the discovery of a coelacanth, and again the press descended on an exhausted and emotionally wrought-up Smith, who rose eloquently to the occasion with an impromptu radio speech. Smith presented a scale from the fish to Malan, who was (as the world's newspapers were quick to note) a former minister in the Dutch Reformed Church, which rejects the theory of evolution. But there was something very wrong; the pilot's joke had been prophetic. The French were upset.

The Comores were then French territory. Smith had landed in a military plane on French territory and taken away a speci-

FIGURE 11   Smith and Courtenay-Latimer examine the second coelacanth soon after it was brought back from the Comores, January 1953. COURTESY OF J. L. B. SMITH INSTITUTE OF ICHTHYOLOGY

men of major scientific importance that had been caught in French waters by a French subject. No wonder they were upset. No matter how much one's heart goes out to Smith, who had made that first leap of imagination and who had labored so long, legally the French (to say nothing of the Comorans) had a right to the fish. Just consider what would be the response of the U.S. general public and U.S. scientists if a live pterodactyl were to be discovered in the Florida Everglades and employees of a French or South African zoo were to fly in, catch it, then parade it as their own. But for the moment there seemed to be no overt response.

Very quickly Smith brought out a short description of the new specimen in *Nature,* with the opinion that it was a new genus and species *(Malania anjouae)* on the basis of differences in the fins from those of *Latimeria:* It lacked the second dorsal fin and the small median element of the trifid tail fin.[47] (Amusingly, the first thing he did in this paper was to take another swipe at E. I. White for his 1939 article in the *Illustrated London News.*) But it was already clear to most zoologists who had seen the photographs that although the specimen differed from the first one, it was really a specimen of *Latimeria chalumnae.* The missing fins noted by Hunt were most probably due to its having been injured, bitten by a shark perhaps, or else the malformation was a natural defect.[48] Only Smith ever really believed that it was a different genus and species. This was the first sour note and harbinger of worse to come.

Even as he prepared his brief description of the specimen for *Nature,* Smith knew that everything had changed with this second discovery. If the true home of the coelacanth(s) now had been found and more specimens would be available, he did not have the resources or breadth of expertise to make the sort of really detailed analyses of the fish that it would need all by himself. Knowing that coelacanths could be caught off the Comores, scientists could go back and get other specimens, and then panels of experts would be needed, each to work on the aspect of the fish that was his or her particular specialty. In

fact, these experts were already bombarding him with requests
for information and material.

So Smith organized a committee under the aegis of the
Council for Scientific and Industrial Research and sent a notice to *Nature* requesting that interested parties get in touch
with him for the purposes of organizing a coordinated research effort.

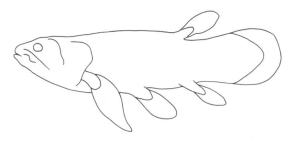

FIGURE 12   Sketch of the second specimen, called *Malania anjouae* by
Smith, showing the missing first dorsal fin and terminal tail lobe, now
thought to have been caused by some prior injury to the fish.

Once again plans were discussed for a big expedition, and
Smith started the process of getting permission from the
French to fish in their waters. And once again the plans fell
through, although a couple of small private international
groups immediately set out to fish for coelacanths.

A team of Italian scuba divers went to the Comores with a
chartered vessel and apparently worked off Dzaoudzi,
Mayotte. They could not dive deep and were, from everything
we now know, in the wrong place.[49] However, eventually a
photograph, claimed to be of a coelacanth, was published.
Most authorities have dismissed this as a fake, and the image
certainly appears to have been retouched. This was the first of
a number of claims of sightings that could not be substan-
tiated, and I suspect that the prospect of foreign groups coming to the Comores like this only contributed to the difficulties

that more serious expeditions would soon encounter in trying to get permission to visit the islands in search of coelacanths.

Dr. Jacques Millot, who was mentioned above as the distributor of the affidavit by Affane Mohamed, was at that time head (and founder) of the Institut de Réchèrche Scientifique in Madagascar (the group in Tananarive with which Coudert had failed to communicate on that Christmas Day). Millot corresponded with Smith about the possibility of the "Scientific Council for Africa" taking over the organization of proposed expeditions.[50] This council was a paper organization for coordinating research in sub-Saharan Africa, and a meeting was organized in Nairobi for October 1953 ostensibly to set up a coelacanth research program. However, by February 1953 the French had already decided to forbid the export of "material of scientific value including coelacanths."[51] In January 1953 Millot had already set in motion an intense fishing effort among the native fishermen of the Comores under the direction of a colleague, Pierre Fourmanoir.[52] The strategy succeeded brilliantly. Just before the Nairobi meeting convened, news came to Smith (typically, he was back in the field at the time, this time in Mozambique) that another specimen, the third, had been taken on September 26, 1953. It was caught off Anjouan again, by a fisherman from Mutsamudu.[53] So it seemed to be certain that the Comores were indeed the home of the coelacanths.

The council meeting came to nothing. Instead of the discovery of a third specimen catalyzing a cooperative effort, it effectively eliminated the chances for any joint work. The French scientists had clearly taken things over, and on November 9, the following press release came from the French: "Only French scientists will be allowed to search for the coelacanth off the French Comore Islands, in the Indian Ocean between Mozambique and Madagascar, for the rest of this year. French authorities there have declared a complete ban until December 31st on expeditions by foreign scientists. . . ."[54] That ban was never really lifted until French colonial rule ended, some fifteen years later.

The French research effort started at the Research Institute at Tananarive but later moved to a special laboratory set up at the Muséum National d'Histoire Naturelle in Paris. Dr. Millot and his assistant, Dr. Jean Anthony, who were originally experts in spiders and anthropology respectively, were put in charge of a research team to study the coelacanth and specifically to prepare a treatise on its anatomy.[55] Under an intense fishing effort, five more specimens were taken in 1954. Two were from Anjouan Island, and three from fishing villages on Grande Comore Island.

The capture of these early French specimens is quite an amazing story. The first was caught in September 1953 by Houmadi Hassani, who recognized the coelacanth as soon as he got it to the surface. He was fishing quite close to shore and quickly got it to his house, where his wife guarded it while he ran for the local French doctor, Dr. Georges Garrouste, who had been given one of a number of sets of materials for preserving a coelacanth that Dr. Millot had distributed. Dr. Garrouste telephoned the administrator for the island of Anjouan,

FIGURE 13   Professors Millot (left) and Anthony examining one of the French specimens at the museum in Paris. COURTESY OF LABORATOIRE D'ANATOMIE COMPARÉE DU MUSÉUM NATIONAL D'HISTOIRE NATURELLE, PARIS, DR. DANIEL ROBINEAU

André Lehr. Together they collected the fish in Garrouste's ambulance and preserved it by injecting seven gallons of form-aldehyde solution. The fish was completely preserved within three or four hours of capture, and it was flown to Millot in Tananarive the next day.[56]

The fourth specimen (taken on January 25, 1954) was the first to be caught on Grande Comore. It was quickly preserved and crated. "It was very exciting, rushing to preserve it, to build a case, to order a special plane from Madagascar."[57] But there was more to come. According to Maurice Jex, adminis-trator of Grande Comore, "we finished up tired and proud at 4:00 P.M., then a man staggered in with an even bigger coela-canth. We went to work on it, and were loading the two boxes on the plane when a third coelacanth was brought in. We were getting tired of fish."[58] In fact, the official tally by Millot, An-thony, and Daniel Robineau shows only two fishes collected off Grande Comore on January 29, 1954. It would be interesting to know what happened to the third if it existed. By the end of 1956 five more specimens had been taken. All were caught by native fishermen using their traditional techniques.

One specimen, the eighth, taken on November 12, 1954, was caught by a famous Comoran fisherman, Zema Mohamed, who eventually caught no fewer than five coelacanths in fifteen years. This one was particularly important because it was still alive when brought to shore. Following instructions that Millot had given should a specimen be brought in alive, a dinghy was submerged to act as a holding pen, and the fish was carefully observed by the local administrators. Millot flew over from Tananarive just before the coelacanth finally died. It had lived for about seventeen hours after being boated. This was the first time that any coelacanth had been seen alive since the first specimen snapped its jaw shut when Captain Goosen touched it. And it was the first time that the behavior of the coelacanth had been scientifically observed.[59]

So the French had begun a whole program of study of the coelacanth, and Smith was completely cut out. Perhaps it would have been too much to ask, in those old territorial, elbows-out days of the cold war and the loss of French prestige in Algeria and Indochina, for Smith to have been allowed to participate in the growing coelacanth research effort. No one else was included either, of course. It became a French effort. Meanwhile, the rest of the world was then watching South Africa in horror as the policies of apartheid, running exactly opposite to the streams of colonial evolution (and revolution) elsewhere came to full view. The name of Dr. Malan, which Smith had hoped to honor, grew in infamy until it became one of the most loathed and despised names of the 1950s, and the regime it stood for was an outcast.

Hunt's story after the discovery of the second specimen turned out to be a tragic one. Fewer than two weeks after Smith took off from Pamanzi with the second coelacanth, Hunt's schooner was wrecked there in a massive cyclone. He and his crew barely escaped with their lives. Ten years later, having resumed his old trading routes in a elderly two-masted 120-ton schooner, the *Hiariako,* he was wrecked on Geyser Bank. As the explorer Quentin Keynes told me the story, Hunt with twenty-five passengers and crew took to a lifeboat and raft, but only five of them reached Moroni on Grande Comore. Hunt and the others were lost to sharks. As for the schoolmaster Affane Mohamed, he rose to become minister for cultural affairs in a later Comoran government under President Ahmed Abdallah (see Chapter 4).

Smith claimed, in his book *Old Fourlegs,* that the discovery of the third specimen lifted an enormous responsibility from his shoulders, for now he was not the only person to have charge of this fish. But without question he was terribly disappointed that a glamorous and extremely important field of study to which he had single-handedly given birth was essentially barred to him. He could only watch while the French seemed to have access to an unending supply of specimens.

After 1952 Smith toiled away at his sea fishes and in the hopes of finding another specimen of his own. He and Margaret Smith completed a major study of the fishes of the Seychelles, and he pursued his studies of the extraordinary diverse marine fishes of Africa to the end, eventually publishing more than two hundred technical papers on them. But what started out in dazzling imagination ended for him in less than full glory. Marjorie Courtenay-Latimer continued to work at the East London Museum, where the importance of the first specimen was eclipsed by the new discoveries. The French group continued to get more specimens and even gave some away. Number fourteen was presented to the British Museum (Natural History) in London and number fifteen to Air Comores for display in Moroni. By 1960 they had twenty specimens, and number twenty-one was given to the Zoological Museum in Copenhagen, where Dr. Eigil Nielsen was one of the world's experts on fossil coelacanths. Specimen twenty-six (1962) went to the American Museum of Natural History in New York.[60] All these transfers were made with the strict injunction that the fish were for exhibit purposes and were not to be dissected for research. As we shall see, this prohibition turned out to have ironic consequences.

Eventually it become possible for others than French scientists to obtain specimens of the coelacanth for research. J. L. B. Smith was offered the chance to buy number twenty, the big specimen that went to New York in 1962, but he turned it down. This is puzzling, but the explanation later offered by Mrs. Smith may be correct: By then too much work had been done by the French team, and the study of coelacanths had long since passed from the sort of work that Smith was skilled at—basic identification and classification and natural history of fishes—to teams of experts in anatomy, microscopic histology, and biochemistry of tissues. The transaction might have been practically impossible anyway. It happened that very early Dr. Bobb Schaeffer of the American Museum of Natural History in New York had offered to assist Smith in assembling a team of ichthyologists to work on a *Latimeria* and even to provide laboratory facilities. When this new specimen (number twenty) was offered to Smith, he suggested that it be offered instead to Schaeffer.[61] The source of this specimen was not the Comoran government or the French research team, but rather none other than Dr. Garrouste from Anjouan Island, who had evidently obtained the fish from a local fisherman and now planned to sell it. This may, in fact, have been the first specimen to be offered for sale privately. Although Smith could not have known it, this big female coelacanth, which did indeed find its way to the American Museum, turned out to be extraordinarily important (see Chapter 9).

In any case, Smith's health was now failing. He had long been suffering from cancer. In 1968, typically tough and resolute in his determination not to risk a stroke or anything else that would make him incapacitated and dependent on others, he took his own life. Indeed, legend has it that he had announced many years earlier that he had no intention of living past seventy. While his widow, Margaret Macdonald Smith, continued the work he started in South African ichthyology and helped build the J. L. B. Smith laboratory at Rhodes University into a major research institution, the center of work on the coelacanth was now Paris. It was to remain in Paris until political events, rather than scientific considerations, forced the opening of the study of *Latimeria chalumnae* to the whole world.

CHAPTER 3

A Living Fossil

The coelacanths have changed very
little since their first known
appearance in the Upper Devonian.
                        —*A. Smith Woodward*

———————

*L*atimeria caught the imagina-
tion of the public. Here was a fish that should have been extinct, a relic from the days of the dinosaurs. Having it in our
world opened up imaginations to the possibility that other fabulous creatures were living in the seas, especially in their most
inaccessible depths. At a time when science seemed to have all
the answers and threatened to take all the mystery out of life,
Latimeria made zoology romantic again and science the realm
of real people, like fishermen and small-town museum cura-
tors.

To zoologists and paleontologists *Latimeria chalumnae,* the living coelacanth fish, is an important element in the jigsaw puzzle of vertebrate biology. The fish is almost always referred to as a living fossil. But what is a living fossil, and just why is *Latimeria* so important?

Zoologists and paleontologists argue a lot about how to define the term and, indeed, about whether it has any meaning at all. It is one of those terms that seem perfectly descriptive and apt to the layman but whose precise meaning for the expert is elusive. Many biologists dislike it (although their most intense feelings are usually directed against its sister term, *missing link*). *Living fossil,* like *missing link,* is an oxymoron—a seemingly self-contradictory construction. (Most people's favorite example of an oxymoron is the cynical one *military intelligence,* another is the description *pretty ugly.*) Obviously, if an organism is living, it cannot be a fossil. But equally obviously, organisms that we know as fossils could also still be alive, unless we mean *fossil* to refer solely to things that are both dug up (the technical meaning of the Latin word *fossilis) and* extinct. That would indeed be silly. So *living fossil* is probably a technically appropriate term, and it is certainly a useful one.

Charles Darwin thought it was useful. He coined the term *living fossil* in his famous book *The Origin of Species* (1859), when he used it to describe primitive living organisms like the lungfishes, which he saw as relics from ancient diversifications, restricted to particular environments where "competition . . . will have been less severe than elsewhere"—freshwater basins, for example.[62]

My best definition is the following: A living fossil is the living representative of an ancient group of organisms that is expected to be extinct (it may for a long time have been thought to be extinct) but isn't. Usually this means, in addition, that the living representative is rare or at least uncommon and has a restricted geographical range. It is a member of a group that was formerly widely distributed in time and space, as indicated in the fossil record, and that otherwise became extinct, usually some long time ago. Finally, there is also usually the connotation that the living representative is itself very primitive in comparison with other groups of organisms, even closely related ones. Any organism that fits this set of descriptions is bound to be interesting to biologists. For example, it causes us to ask, How and why has this species survived when all its relatives died out? What does it tell us about patterns and rates of evolution in the fossil and living record? What do living fossils tell us about the groups that have not survived and about the groups that have continued to flourish?

A prime example of a living fossil is the horseshoe crab, restricted to a group of four or five species, the best known of which is *Limulus polyphemus,* the common horseshoe crab of the eastern coast of the United States. Horseshoe crabs are the sole representatives of a large group of fossil invertebrates called Xiphosurida, more closely related to the spiders than to crabs. The first members of this group appeared some 425 million years ago in the Silurian and look quite like the modern forms. The last fossils became extinct about 50 million years ago. Interestingly, the species *L. polyphemus* is not rare. Millions of individuals occur in populations on the western North Atlantic, and they are famous for their mass spawning on the beaches of Cape May or Chesapeake Bay and for the associated feeding frenzy of seabirds that swoop down to prey on the eggs from the moment of release by the females.

A second classic example of a living fossil is the dawn redwood tree, *Metasequoia glyptostroboides,* a conifer related to the various Pacific coast redwoods of North America. This species was widespread and reasonably common in the Pliocene of North America. Pollen spores of this tree regularly turn up in fossil samples, and indeed, these fossils were formerly its only known record. The tree was thought to be extinct worldwide until living specimens were found in central China in 1945. A nice puzzle here is that when the species was reintroduced to North America, it turned out to do quite well; handsome stands may now be seen in botanical gardens and arboreta such as the Morris Arboretum in Philadelphia. So why did *Metasequoia glyptostroboides* die out here? Probably the Ice Age glaciations temporarily obliterated those habitats where it could survive in North America.

The Virginia opossum, *Didelphis virginianum,* which has become a common urban pest of America's East Coast, is a third example—the only marsupial in North America. Left over from a once-wide radiation in the New World of marsupial mammals related to the kangaroos, opossums, and wombats, this opossum is related to organisms for which Australia and New Guinea are now the principal homes. Like *Metasequoia,* the opossum historically had a restricted distribution but has recently spread widely. Its rapid spread northward in the last seventy-five years is probably due both to a climatic warming and to a new and abundant supply of food in the form of garbage.

The term *living fossil* is sometimes also used to label primitive living forms that have no known fossil relatives at all, or at least no close relatives of the same general sort, but are so primitive that they must be relics of some ancient diversification that has long ago disappeared from the face of the earth. For example, in 1955 Dr. Howard Sanders, then a graduate student at Yale, found a strange and primitive-looking arthropod in the ooze at the bottom of Long Island Sound, of all unglamorous places.[63] This arthropod turned out to be a completely new and very ancient type of arthropod—the Cephalocarida—with no living or fossil relatives. This form is obviously left over from the Cambrian (perhaps five hundred million years ago), when complex living crustaceanlike organisms were just beginning to evolve on earth.

Several sorts of fishes could be called living fossils. All three genera of living lungfishes—relatives of the coelacanths—fit in this category. The lungfish, or Dipnoi, arose in the Early Devonian (some 375 million years ago), and the group diversified widely. By 200 million years ago they were already reduced to a very few forms, but three genera survive today. All

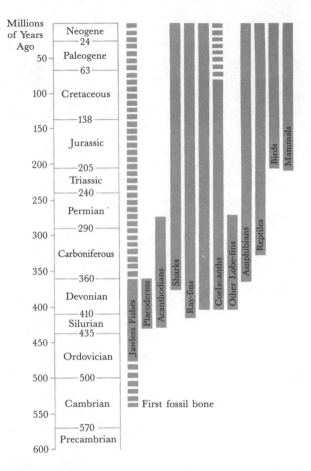

FIGURE 14   Geological time scale showing point of origins of major vertebrate groups.

have a restricted geographical distribution, and all live only in freshwater, although many of the earlier fossils lived in salt water. The Australian lungfish, *Neoceratodus forsteri,* preserves the most primitive morphology, with large lobelike fins—the sort of fin that tipped off Marjorie Courtenay-Latimer that the coelacanth was something ancient and important. It is restricted to two river systems in Queensland on the east coast of Australia, although it apparently has flourished in artificial lakes and reservoirs in the last 50 years. The African and South American lungfishes are more advanced, showing highly specialized fins, reduced to very sensitive feelers. The South American lungfish *Lepidosiren paradoxa* is restricted to the Amazon Basin. The African lungfish genus *Protopterus* is the most diverse of the three types and is widespread in tropical Africa. Fossil lungfishes have been found on every continent, and the far-flung distribution of the living forms reflects the drifting apart of the southern continents on which they were trapped during the late Mesozoic and their exclusion from other environments by more modern kinds of fishes.[64]

The Mississippi River has several "living fossil" fishes: the bowfin *(Amia calva)*, the paddlefish *(Polyodon)*, the sturgeons, and the gars *(Lepisosteus* and *Atractosteus)*. This great river system represents an extremely ancient drainage basin, and these fishes have descended from ancient lineages. They are important fishes to zoologists because they preserve the features of fishes from key episodes of change in the evolution of the great radiations we call ray-finned fishes, which make up most of the living fishes like cod, perch, tuna, etc. The gars, for example, still have heavy, shiny scales of the ganoid type that were typical of fishes living between the Devonian and Triassic, related distantly to the sorts of scales ancient lungfishes and coelacanths had.

Measured against these standards, the coelacanth *Latimeria chalumnae* fits the category of "living fossil" quite well. There is only a single living species. It belongs to an ancient, formerly more diverse group. The last known relatives of the living coelacanth became extinct in Cretaceous time, some 80 million years ago. The fish is morphologically very primitive in that its structure seems almost identical with what we can reconstruct of the fossil forms, even the earliest fossils that are Late Devonian in age (some 375 million years ago). It seems to have a highly restricted distribution.

Organisms fitting the category of "living fossil" are the

more important to us, the more conservative their evolutionary history has been. In the coelacanth *Latimeria* we see preserved a skeletal morphology essentially identical with that of
Devonian coelacanths. Therefore, we argue that if the skeleton
has not evolved significantly since Devonian times, the rest of
its general biology—its physiology and reproduction, for example—may be equally primitive. In that case, by studying the
living fossils, we can get a genuine firsthand glimpse of
Devonian, or at least very ancient, biology and of physiological
features lost in modern fishes. Just to have the chance to compare the full anatomical structure of a living fossil—both the
hard and soft parts—with the features that are available in fossil representatives gives us a wonderful new perspective on the
interpretation of the fossil forms. Where we find muscle scars
on the bones of a fossil, for example, direct evidence from the
living form will allow us to tell what those muscles must have
been, and from that we can reconstruct how the fossil animal
must have moved or fed.

For all these reasons, then, living fossils like *Latimeria chalumnae* are extremely exciting to zoologists. In addition, of
course, one should not underestimate the energizing effect
that any glamorous discovery has on a field of study, just
through making it fashionable and popular. It is interesting to
speculate, for example, whether coelacanths would be so interesting to the public and ultimately so popular among scientists if *Latimeria chalumnae* were a common fish of North America or Europe, or if it had been only six inches long instead of
five to six feet. There is something particularly fascinating in
the fact that so large a fish could have remained unknown to
science for so long.

Perhaps, by studying a living fossil like *Latimeria chalumnae*,
we can begin to explain just how certain species managed to
survive as such unique cases. Is there something special about
the organism itself, something about its environment? Or is all
simply chance? Of what significance is the fact that many living
fossil forms that seem to be primitive relics actually flourish

when given the right circumstances? Does the survival of living fossils represent the operation of some common factor, or does each species survive for a unique reason?

If all this were not enough, as we will discuss in a moment, the coelacanths, living and fossil, are particularly interesting to zoologists because of their position in the great scheme of relationships of organisms—the phylogeny of animals. The coelacanths belong to the group of fishes from which, way back in Devonian times, the first ancestors of land vertebrates arose. Coelacanths are related to the ancestors of amphibians and all other land animals—the reptiles, birds, and mammals—and therefore us.

Latimeria chalumnae was recognizable to Smith because he knew the fossil record of fishes and knew the similarity between this new fish and coelacanths that had been studied as fossils long before. Coelacanths had been known for almost exactly one hundred years before Smith saw Courtenay-Latimer's sketch. In 1836 no less a zoologist and paleontologist than the great Louis Agassiz (several years before he left Switzerland for America) described a fossil fish of Permian age from Ferry Hill, near Newcastle, England, in his classic book *Poissons Fossiles.* [65] He gave it the names *Coelacanthus* because the spines of the first dorsal fin were hollow (*coel,* Greek for "space"; *acanthus,* "spine") and *granulatus* for the tubercular ornamentation of the surface of its scales. Technically the full name of the species was *Coelacanthus granulatus* Agassiz 1836. This was the very first coelacanth to be described and gives the order Coelacanthini its name.

In zoology and botany names are vitally important, so at this point it is necessary to mention a matter of biological legalistics. Each distinct kind, or "species," of organism must have its own name, simply so that every zoologist will know what anyone else is talking about. The first comprehensive scheme of nomenclature was founded in 1758 by Karl von Linné (Linnaeus) in Sweden, both to formalize naming systems and to allow scholars to fit each species into schemes of relationship. Each species has its own name, and this is always expressed as a binomial plus the name of the scientist who first formalized the name. Our own binomial is *Homo sapiens* Linnaeus. *Homo* is a name in the category "genus" and *sapiens* in the category "species." One reason for the binomial is that like the letters and numbers on car license plates, it gives more combinations. Many organisms have the species name *vulgaris* or *domesticus,* for example. But there is only one *Passer domesticus* (the common European house sparrow) or *Sternus vulgaris* (the European starling), and so on. The rules of Linnaean naming are simple: Generic names must be unique, and the species names must not be used more than once per genus.

The generic category also starts the process of organizing the names and the organisms. *Felis catus* is the house cat; *Felis lynx* is the lynx; *Felis concolor* is the North American mountain lion. The fact that all are named to the genus *Felis* indicates that zoologists think that all cats are closely related—not closely enough all to be *catus* or *concolor,* but more closely than to be members of the genus *Panthera* (which is the taxon for the lion, *P. leo*). There are many levels of category beyond the genus. The next higher level after genus is family; the family Felidae includes both *Felis* and *Panthera,* for example. And above family is order; the family Felidae is one of many families: Canidae (dogs), Ursidae (bears), and so on, in the order Carnivora.

One point has to be emphasized; The living coelacanth is not a living fossil in the very strict sense that members of the species *L. chalumnae* itself have ever been found as a fossil. In fact, no other species assignable to the genus *Latimeria* has been found as a fossil either. *Latimeria* and the Cretaceous fossil genus *Macropoma* are quite closely related, and we could possibly include them in the same family. Beyond that, all fossil coelacanths belong in the order Coelacanthini (which a minority of zoologists would prefer to call Actinistia, a name for which I can find no use at all).

THE FOSSIL RECORD

The most ancient coelacanth that anyone knows is *Diplocercides,* which is Late Devonian in age.[66] The group as a whole is therefore at least 375 million years old and probably older. We can see in *Diplocercides* an extraordinary number of the features of *Latimeria chalumnae* (features of the skeleton, of course, because that is what gets fossilized). For example, the first coelacanth certainly had the same rostral organ, intracranial joint, paired fins, vertebral column, hollow notochord, and reduced teeth. This tells us that the group as a whole has not evolved much since the Devonian, but it also tells us that there is a big gap in the record: We are missing the sequence of even older ancestral fossils that should one day show us how the particular suites of characters that are common to all coelacanths arose. We do not yet know what the immediate ancestor of all the coelacanths was.

The first evidence of fishlike vertebrates is small fragments of bone from the Cambrian of North America (more than five hundred million years ago). Then there is a great gap until the mid-Ordovician, when remains of archaic fishes are found in a great chain of deposits along the Rocky Mountains. These fishes belonged to an early radiation of vertebrates that had a somewhat simple mouth structure and feeding mechanism. They lacked teeth and the sort of specialized jaws that typify more advanced fishes. Instead, they had a sort of sucking mouth and may have fed by slurping up detritus and small organisms on the bottom. These fishes seem primitive to us; but their radiations lasted well into the Devonian, and in evolutionary terms they were very successful.

Two sorts of living fishes (living fossils once again) are descended from these archaic forms. These are the lampreys (parasites on other fishes and famous for their depredations on the fishing industry of the Great Lakes) and the hagfishes,

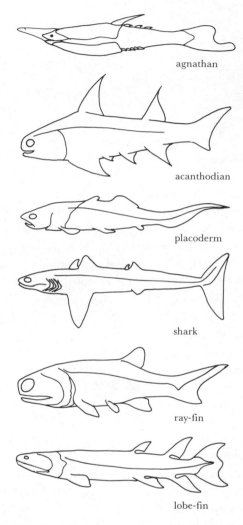

agnathan

acanthodian

placoderm

shark

ray-fin

lobe-fin

FIGURE 15   Sketches of representatives of the major groups of fossil fishes.

denizens of the sea bottom that scavenge carcasses of dead organisms. Both are eel-shaped and have no jaws and no fins.

Not until the Late Silurian do we pick up in the fossil record evidence of the origin and radiation of a more modern sort of vertebrate—one with jaws and teeth and a full complement of fins, hence the ancestors of all modern vertebrates. As is always the case with the fossil record, the key ancestral forms are missing, but by the end of the Silurian we can already find the first examples of all four of the major groups of jawed vertebrates, or Gnathostomata (*gnathos,* "jaw"; *stoma,* "mouth"). These were the Acanthodii (a group of mostly small fishes, perhaps planktonic filter feeders, that was extinct by the mid-Permian); the Placodermi (a group of heavily armored fishes that radiated very widely in the Silurian and Devonian and gave rise to some monster predators up to eight feet long, before becoming extinct at the beginning of the Carboniferous); the Chondrichthyes (the sharks and rays and their relatives); and the Osteichthyes—the so-called bony fishes (poorly named because other groups are also bony)—which included two groups that we distinguish on the basis of their paired fins as the ray-fins (Actinopterygii) and the lobe-fins (Sarcopterygii). The bony fish, or Osteichthyes, comprises the great majority of living fishes, but of course, only four genera of these are of the lobe-finned sort—namely, the three lungfishes and *Latimeria.*

Chondrichthyes — sharks, skates, rays

Osteichthyes — all bony fishes
  Dipnoi — lungfishes
  Crossopterygii — fringe-finned fishes
    Coelacanthini — coelacanths
    Rhipidistia — no common name, extinct
  Actinopterygii — ray-finned fishes

FIGURE 16 Scheme of relationships of modern fish groups current in Smith's time.

Of the early radiation of fishes it is fascinating that so many groups failed to make it past the Paleozoic (or even past the Devonian), while the sharks and rays and the ray-finned bony fishes not only survived and replaced them but flourished in the Mesozoic and Cenozoic at levels of diversity far beyond anything seen in the Paleozoic. The sharks and rays are key predators in modern seas but only very occasionally throughout their whole history did the group penetrate into freshwater. The ray-fins got the slowest start of all—their Silurian and Devonian record is very sparse—but once they reached the Carboniferous, they started to diversify, and then wave after wave of new groups appeared and continues to appear. We find relic species, left over from the early phases of all these radiations, persisting today as living fossils, which help enormously in trying to understand their evolution. There are several living fossil sharks, and among the ray-finned fishes there are the sturgeons, the bowfin *Amia,* the African bichirs *Polypterus,* the gars *Lepisosteus and Atractosteus,* and the paddlefish *Polyodon,* mentioned previously.

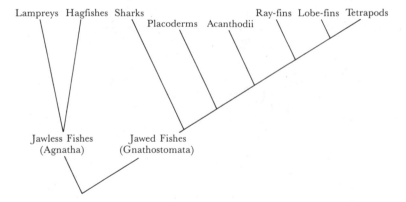

FIGURE 17 Modern view of the relationships of the major fish groups.

Coelacanths belong to the osteichthyan (bony fish) group called lobe-finned fishes or Sarcopterygii (*sarcos*, "fleshy"; *pterygium* "wing" or "fin"). As the name suggests, the group is characterized by a particular sort of fin structure that is not found in other groups of vertebrates. The distinction is seen most clearly in a comparison between a lobe-finned fish like *Latimeria* and a fish like a trout that belongs to the other principal group of living fishes, or Actinopterygii—the ray-finned fishes.

In the ray-finned fishes (perch, salmon, goldfish, eel, piranha, tuna)—essentially all the fishes we know that are not lobe-fins or sharks and rays—with only a few exceptions, there is only one dorsal fin. Both the median and paired fins always have the same basic pattern of structure, each consisting of a number of bony, flexible fin rays, inserted directly into the body wall, with a web of skin stretched across them. The fin rays can flex slightly, and the rays can be moved individually by muscles within the body wall, so as to "fan" or "scull" the fin. In lobe-fins, by contrast, the second dorsal, anal, and paired fins are of a quite different type in which there is a stout muscular stomp or "lobe" projecting from the body wall. This lobe has its own internal skeleton and muscles, and the fin rays are restricted to a fan that attaches to the outer end of this stump. The fin rays have all the mobility of the fin rays in the ray-finned fishes, but in addition, the fleshy scale-covered lobe is flexible and can be moved (bent or rotated) both by its own internal musculature and by muscles running from it to the body wall. The result is a much stronger, more mobile flexible and adaptable limb.

Is one sort of fin *better* than another? Probably not. They are merely different and good for different things. Hindsight suggests to us that the lobed type of fin was good for pushing against the bottom and perhaps, in shallow water and marshes, for lifting the head out of the water and supporting the front part of the body so as to allow the fish to breathe. From this it was a short step to pushing the animal around in the shallows

FIGURE 18   The key features of the "lobed" paired and median fins of *Latimeria,* compared with the fins of a carp (below). Internal bones of the lobed fins are shaded.

and then to moving on land. But the lobed-fin condition must also have been perfectly good for swimming around in the water without touching the bottom at all; the jointed internal structure is suited for making complex sculling actions. We do know, however, that the ray-finned fishes diversified enormously while the lobe-fins were becoming extinct. In evolution the ray-fins have given rise only to other ray-fins—but to an incredible diversity of ray-fins. The lobe-fins gave rise to the tetrapods in the Devonian, and most of them soon became extinct not long thereafter.

It is easy to conclude from at the history of all these groups, that the lobe-fins were driven to extinction or near extinction by competition with the ray-fins and sharks in the water and by competition with their descendants the Amphibia in the marshes and on land.

In the Devonian, coelacanths were fairly common but not really abundant; some fifteen to twenty genera are known so far, and at least twenty different species (to distinguish species among fossils in the same terms as one distinguishes species of living fishes, when one lacks data on color patterns, behavior, etc. is very tricky, if not impossible).* The group was about equally abundant in the Carboniferous, but the Devonian species were already extinct. By the Permian the whole group was almost extinct. Curiously we then find an excellent record of Triassic coelacanths, some thirty species in all. Prominent among these was the genus *Diplurus,* which was widespread in Europe, North America, and possibly China. Thousands of specimens have been preserved in the fossilized remains of lake beds in the northeastern United States. This fact serves to remind us that our knowledge of the distribution of fossils is never any better than the rocks themselves allow. Where a group seems extinct, it may merely be that we haven't found the right types of rocks in which their fossils might be preserved. In the Jurassic the group was down to fifteen or sixteen species, and in the Cretaceous there were fewer than ten, and there the story was thought to end. The youngest fossil coelacanth is in the genus *Macropoma,* found in the Late Cretaceous of Europe and Asia.

A note must be added here about the famous Cretaceous extinctions, not just of dinosaurs and other vertebrates but of many life-forms. A theory has recently been widely popularized that the extinctions at the end of the Cretaceous were caused by the impact of a very large asteroid with the earth.[68] There is a great deal of geological evidence for such an impact, and if it happened, its effect must have been virtually instantaneous. To grasp the immediacy of the phenomenon, I find it useful to keep it firmly in mind that it all had to happen on a Tuesday afternoon and by two Sundays later the environment of the earth would have been significantly altered. The trouble is, when one looks at the fossil record of organisms living (and

---

*There are always surprises in the fossil record. In the last few years a new, totally unsuspected fauna of fishes, including several new species of coelacanths, was discovered in the Upper Devonian of Montana.[67]

dying) in the Late Cretaceous, it becomes clear that they declined at different times and over periods of time ranging from
several thousand to several million years. Many of these de-
clines were under way long before that fateful Tuesday (or
perhaps Friday) afternoon. The last fossil coelacanth record is
some fifteen million years before impact, for example. Many of
the effects of the impact were soon reversed. Therefore, while
there probably was an impact and it probably did have a major
effect, it almost certainly does not explain the extinction of
dinosaurs, coelacanths, or many other groups. Conversely,
other groups, such as birds and modern fishes, sailed through
the apparent devastation of the earth at the end of the Cretaceous as though nothing had happened at all.

   When we look at the radiations of coelacanths in the fossil
record, several important features are immediately striking.[69]
First, the group is an extremely ancient one in comparison
with other groups of fishes, yet at no time was there ever more
than a handful of species of genera around at one time; their
taxonomic diversity has always been very low. At the same time
some genera and species have individually been extremely nu-
merous and widespread. In cases like this paleontologists
would give a lot to know how much of the fluctuation in diversity is real and how much is due to unreliability of the fossil
record itself. We often do not know whether a shortage of
fossils means that environments that were congenial to coelacanths and in which they might have been incorporated into
sedimentary deposits as fossils were rare or abundant or
whether we simply see differential persistence of those rocks to
the present time. Perhaps there were once lots of coelacanths
that we do not know about. Perhaps lots of them were fossilized but the rocks containing them were ground up and swept
away by glaciers or eroded by rivers aeons ago. Or perhaps
they were buried deeper in the earth by later strata so that we
have not yet found them. Each year that paleontologists explore the surface of the earth, more and more fossils are
found, and equally, each year that we fail to find the "missing"

diversity of coelacanths, the more the low diversity that we observe tends to seem real.

The great majority of fossil coelacanths were marine or brackish water fishes. Only a few are found in strata and sediments that indicate freshwater environments. Nearly all the Devonian forms are marine. In the Carboniferous (Mississippian and Pennsylvanian), where there happens also to be a great abundance of swampy brackish and estuarine environments preserved in the fossil record, we find coelacanths that must have been able to tolerate a full range of environmental conditions from salt to partially freshwater. *Rhabdoderma,* a smallish coelacanth, the size of a large minnow, is quite common in coal deposits of both Europe and North America. In the Late Triassic the extremely abundant genus *Diplurus* mentioned above was definitely living in freshwater lakes and rivers of North America. Also, up to this time almost all fossil coelacanths had been small fishes of less than eight to ten inches), but one species of *Diplurus* was much bigger (to fifteen inches).

In the Jurassic we find both freshwater and marine forms, and some species were quite large. In the Cretaceous, on the other hand, we find more marine forms, particularly in the great chalk deposits of Europe, although curiously they are absent from the comparable chalk deposits of North America.

Apart from the Late Triassic/Early Jurassic lake deposits of eastern North America, the region that has produced the most abundant record of coelacanth fossils has been Madagascar. In the Lower Triassic marine strata of Madagascar there is an enormously productive series of fossil beds principally yielding fishes and ammonites, indicating a shallow warm-water environment.[70] Among the more than thirty species of fishes represented by thousands of individual specimens are at least four species of coelacanths.

The last (that is, the youngest) fossil coelacanths that we know are marine forms from the Cretaceous period. *Macropoma,* a fish of about thirty centimeters from eighty-million-year-old Late Cretaceous chalk deposits of Europe, seems to have been the last coelacanth to be preserved as a fossil. Apart from size, it is very much like *Latimeria.* A somewhat earlier fish, from the Early Cretaceous of Brazil, was *Axelrodichthys,* recently described and named by Dr. John Maisey of the American Museum of Natural History.[71] It was very large, at least as large as *Latimeria.*

The fact that we have no fossil coelacanths from strata younger than the Late Cretaceous is probably explainable in terms of both the actual distribution of coelacanths during that time and the conditions of preservation. First, it is probable that whatever coelacanths existed though the Cenozoic period were marine forms. Second, we have only a modest record of marine fishes from the Cenozoic and a very poor record of marine fishes from deeper waters than those that inhabited the shallow continental shelves. Third, as we will discuss further below, it is possible that coelacanths in the Cenozoic happened to live in a region—namely, the western Indian Ocean—from which we have almost no suitable marine fossil record at all.

It is interesting that a group unique in terms of skeletal and other adaptations should always have been so sparse in terms of numerical diversity of genera and species. For evolutionary theorists this raises the question of the extent to which the progress of evolutionary change requires the production of a great diversity of species at any one time. Numerically the coelacanths are only a minor branch in fish evolution; today they are represented by only one—*Latimeria chalumnae*—of perhaps more than thirty thousand living species of fishes. Morphologically the group has been conservative in its evolution. But there is another surprise in the fossil record. In the Triassic of Greenland there is a coelacanth genus *Laugia* that has a remarkable set of adaptation. Its hind or pelvic fins have become moved all the way forward and connect with the shoulder girdle; the pectoral fins accordingly have moved dorsally.[72] This change gives a whole new function to the fins as braking, steering, and swimming devices and would be interesting enough in its own right, but it happens also to be an identical type of adaptation to one that is found in the most advanced groups of ray-finned fishes. Unless the fossils have been misinterpreted, and that seems very unlikely, this is a superb case of convergent evolution, and it requires us to hesitate before we call the coelacanths, or any other group, conservative or primitive.

Even before we had the bench mark of *Latimeria* by which to interpret the fossil forms, paleontologists were interested in reconstructing as much as possible concerning the biology of coelacanths. Their fossil record gives many clues but presents many puzzles to be explained. All known coelacanths lack the principal upper jawbones (maxillae) and their teeth, and all have the curious intracranial joint, which seems to indicate a specialized feeding mechanism. They all have the curious rostral sense organ, the same fin and tail structures, and all seem to have lacked a bony backbone.

Because of the relationship of coelacanths to the lungfishes, paleontologists have been particularly interested in the question of whether they had lungs and, indeed, when lungs first arose. At one time lungs were thought to be characteristic pri-

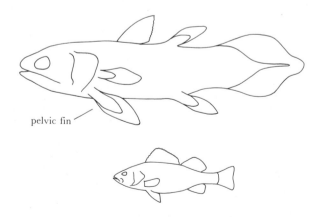

pelvic fin

FIGURE 19 The Triassic coelacanth *Laugia* showing the changed position of the pectoral and pelvic fins, convergent with a modern perch (below).

marily of freshwater fishes, and marine fishes like coelacanths therefore might not be expected to have them. But in many fossil coelacanths, for example, in the well-preserved fossils of *Macropoma* from the (Upper Cretaceous) English Chalk, *Axelrodichthys* from the Santana Formation of the Lower Cretaceous of Brazil, or the Coal Measures *Rhabdoderma,* one can see perfectly well a large organ that must be a lung or a derivative of the lung. It is single and median rather than paired, but that is quite common as a specialized condition in other fishes. More surprisingly, this "lung," in several fossil forms ranging back to the coal-shale fishes of the Carboniferous, has a wall that is armored with tiny bony scales. This suggests that the "lung" was not flexible and could not fill with air and empty. When Smith found a remnant of the lung in *Latimeria,* he identified it by comparison to this median structure in the fossils and by comparison of both to the median lunglike structures of other fishes. It was identical except that it lacked the scales.

The fossil record has also given us lots of puzzles with respect to the reproduction of coelacanths. Most fishes are technically oviparous. That is to say, they *lay* their eggs. They may lay them on the bottom, or they may discharge them into the water to float among the currents. Usually fertilization of the eggs by the male occurs in the water at the same time that the female sheds her eggs. However, in a surprising number of cases, especially in sharks and rays and some very advanced modern fishes (including many of the "tropical fish" that we keep in home aquaria), the young are born alive. For this to occur there must be internal fertilization of the eggs within the female's body, and then the eggs must be retained within the mother's body until the hatchling stage before being shed. These fishes, like some reptiles (particularly snakes), are then said to be ovoviviparous (egg-live-bearing). Advanced mammals, like humans, are truly viviparous (live-bearing) in that they have dispensed with the specializations of the eggs that make them self-sustaining while within the female (the eggs merely being protected by being inside her body) and actually nurture the embryo directly through a special organ, the placenta, rather than through yolk stored in the egg.

There is one genus of fossil coelacanth—*Undina* (also called *Holophagus*)—from the Jurassic of Europe, in which very fine details of internal structure of fossils are often preserved. These specimens come from the famous lithographic limestone locality at Solnhofen, Germany, from which *Archaeopteryx,* the first primitive fossil bird, was found. Professor D. M. S. Watson, the London paleontologist whose name has already been mentioned, discovered specimens in 1926 that show a fish with two miniature coelacanths inside it.[73] He identified them as embryos lying within the reproductive tract of a female. If this interpretation is correct, *Undina* is a live-bearer. (The only alternative is that this is a case of cannibalism!) But these tiny specimens are located far back in the body cavity, rather than forward, so they do not seem to represent food items in the stomach. And they are incompletely ossified and certainly the right size to be embryos. Most authorities after Watson accepted his identification of this coelacanth, at least, as a live-bearer. But, as we shall see, when zoologists came to study *Latimeria,* the matter was thrown in some doubt again.

Why else are we so interested in coelacanths, apart from the fact that a living species of an otherwise extinct ancient group has been found?

There were three or four main lineages of the lobe-finned fishes. Zoologists are still arguing about the exact way to divide them up and about their relationships, but they generally agree that they form a single group that probably arose at the end of the Silurian and had its first major radiation into three or four distinct groups by the very early part of the Devonian. One major group, the lungfishes, or Dipnoi, has a record that almost exactly parallels that of coelacanths: a strong Devonian radiation, dwindling by the Permian, some modest Triassic/ Jurassic success, and then extinction except for the three relict genera surviving to the present day. In contrast with *Latimeria,* the three living lungfishes *are* known as fossils from a range of

Cenozoic deposits. Undoubtedly the more continuous fossil record of the lungfishes exists because they were freshwater forms with a habit of burrowing in mud.

Dipnoi are obviously lobe-fins. Like coelacanths, they have a well-developed notochordal portion of the backbone and weak vertebrae, but there are some prominent differences between them and the coelacanths. There was no intracranial joint in the head and no rostral organ. All Dipnoi have a specialized head fused into a solid structure bearing large crushing tooth plates. Dipnoi evidently were (and are) principally bottom feeders, eating shellfishes and other invertebrates.

Apart from the coelacanths and lungfishes, there is another group of lobe-finned fishes, known to us only as fossils and whose relationships are still being debated. These also first appeared in the Devonian, and all had their most broad diversifications in the Devonian. But none of them survived beyond the Early Permian. Formerly these lobe-finned fishes were grouped together as the Rhipidistia—a term that is in wide, though almost certainly incorrect, usage today and probably comprises several totally distinct lineages. The Rhipidistia used to be linked with the Coelacanthini as a larger group Crossopterygii, a grouping that also may no longer be valid.

The lobe-finned fishes formerly called Rhipidistia are elongate predatory fishes with two dorsal fins. As in the coelacanths, the second dorsal fin, the anal fin, and the paired fins are of the special "lobed" sort. All had the intracranial joint, which Dipnoi lacked, so they are thought by most people to have a close relationship to the coelacanths. Like all the lobe-fined fishes, they had a very scanty set of bony structures in the backbone and the notochord was large and strong, extending forward under the head. But there was no rostral organ. Some of the Rhipidistia have a sort of trifid tail, but it is always slightly different from the coelacanth tail.

While zoologists argue about how many separate lines of lobe-finned fishes there were, all agree that from somewhere within this assemblage, no later than the Late Devonian, the first land vertebrates or amphibians arose.

As we will discuss in Chapter 10, the principal reasons for thinking that the lobe-fins were the ancestors of all the tetrapods have to do with extraordinary similarities in the structure of the head and the paired fins. These similarities are so great that when the first living lungfishes were found, they were thought to be amphibians, not fishes. In fact, it is an extraordinary coincidence that one of the most exciting discoveries of a living fossil organism prior to the discovery of *Latimeria* was the discovery of their close cousins the living lungfishes. The first to be found was the South African lungfish *Lepidosiren* in 1836, closely followed by the African *Protopterus,* which was similar enough to *Lepidosiren* to have been put in the same genus at first. Then came the Australian *Neoceratodus* in 1870.[74] This last was especially important because it showed the lobe-fin structure so beautifully. All three were fishes with lungs that could survive out of the water at least for short periods. In fact, we now know that the gills of the African and South American species are so reduced that these fishes will actually drown if they are prevented from breathing air. All lungfishes have internal nostrils, somewhat like those of tetrapods, and all had the special limb structure. In fact, with the discovery of the lungfishes, the question of the closest relatives of tetrapods seemed to have been solved.

However, at the same time as the fossil coelacanths were becoming much better known, so were their cousins the Devonian and Carboniferous fossil Rhipidistia. In 1892 Edward Drinker Cope, from the Academy of Natural Sciences in Philadelphia, proposed an ancestry of tetrapods from a group of rhipidistrians called osteolepiforms, displacing a previous theory that the Dipnoi were ancestors.[75]

No one has ever held a theory that coelacanths were the direct ancestors of the tetrapods, merely their first cousins, as it were. But because we have no living osteolepiforms to study (and, in fact, no really primitive living amphibians either), evolutionists who try to reconstruct the sequences of events of the fish to tetrapod transition must work from comparisons of those Devonian fossil forms with the closest living

species that are available—the living lungfishes and the coelacanth.

Some scientists have recently returned to consider the hypothesis that the immediate ancestor of land vertebrates might
have been more like a lungfish than an osteolepiform. We
don't need to borrow that quarrel, inasmuch as the prime subject here is coelacanths. Everyone agrees that coelacanths
were not the immediate ancestor, so they don't have quite the
pride of place that one would like. But they are one of the only
four genera of living lobe-finned fishes to remain out of the
great Devonian radiation of lobe-finned fishes that was crucial
to the matter. And they are particularly important because of
the conservatism of their evolutionary change. If coelacanths
or lungfishes had continued to evolve wildly in a morphological sense through the Mesozoic and Cenozoic as the ray-finned
fishes have, they would not be so useful to us. We cannot take a
perch or a cod as a model for understanding its early Devonian
ray-finned ancestors. But *Latimeria* looks very much like the
Devonian *Diplocercides* or *Nesides*. Therefore, by studying *Latimeria* in detail, alongside the three lungfishes and in conjunc-
tion with the fossils and the physical evidence that the rocks
themselves provide about the ancient environments in which
they lived, we might be able to reconstruct a lot about the
biology of those long-distant Devonian forms. And not just the
evolution of the skeleton, but the blood, the liver, how they
breathed, how they reproduced, how they fed and swam; their
whole biology.

We might also hope to be able to discover something in the
biology of *Latimeria* and the lungfishes that would give us a
clue to why and how four lineages survived while all others
became extinct. The great radiations of ray-finned fishes
mostly occurred after the Cretaceous. No lungfishes survived
in the seas after the Cretaceous, but they did survive in freshwater. What happened to coelacanths at the end of the Cretaceous we cannot tell. No fossils are known, so we must conclude that they did not inhabit the sorts of environments from
which we have Cenozoic fish fossils, which means principally

shallow epicontinental seas and freshwater. There are few deeper sea deposits of Cenozoic age. There are very few marine Cenozoic deposits from around the rim of the Indian Ocean where one could look for coelacanth fossils.

HISTORY OF THE INDIAN OCEAN

The reasons for all this are apparent if one studies the movement of the landmasses that have floated around the surface of the earth throughout the last billion years.

The emergent landmasses of the earth's surface belong to a mosaic of major and minor crustal plates, fifty miles thick or more, that move on the fluid mantle of the earth beneath, imperceptibly but inexorably according to still-unknown rhythms. This phenomenon of plate tectonics results in continental drift, in which the continents have moved relative to each other over time and still continue to move, and in sea-floor spreading, in which magma rises up to form new ocean floor crust at mid-ocean ridges as the plates move apart from each other.[76] The churning movement of the crust is responsible for the opening up of ocean basins, great fault zones like the African Rift Valley, and smaller phenomena like the San Andreas Fault in California (where two sections of plate are moving past each other). All this slow-motion turmoil is difficult to comprehend, because we are used to thinking of the earth beneath our feet as solid and reliable—terra firma—but it is accompanied on a more appreciable scale by earthquakes and volcanic eruptions that are only too real.

As is now familiar, the continental plates have moved around a great deal over the hundreds of millions of years of geological time. In the Devonian the great crustal plates were partially separated from each other. But by the Triassic they had moved together and combined into one great landmass, which paleographers call Pangaea, the reality of which is most obviously reflected in the neat fitting of the present eastern coastline of the Americas into the western coastline of Europe

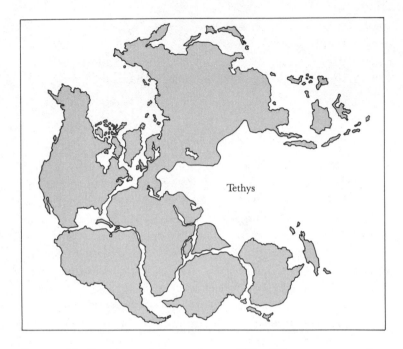

FIGURE 20   Positions of the continents 250 million years ago, when they were united to form a single landmass, Pangaea.

and Africa. Through the late Mesozoic and Cenozoic the plates drifted apart again toward their present (still-changing) positions. It was this process that trapped the living lungfishes in their present disjunct distribution.

On a map of Pangaea for, say, the Triassic, we see that there is no Atlantic Ocean and no Indian Ocean. There is, however a huge ocean, called Tethys, that is open to the east and reaches westward in between the Eur-Asian and African-Arabian coasts. When Pangaea broke up, starting in the late Mesozoic, the continental complexes moved away from each other, opening up the Atlantic and Indian oceans, but they also closed off Tethys, leaving only a remnant in the form (partially) of the present Mediterranean and Black seas.

In the process of the breakup of Pangaea, the surface of the earth was variously stretched and compressed, the plates surging (in a geological sense, at least) into each other and causing the formation of great mountain folds, such as the Himalayas, which are among the world's most recent mountains, rising where the Indian subplate is being forced into Asia. Intense earthquake and volcanic activity is also found around the edges of the major crustal plates as they move past each other; this causes the so-called ring of fire, the zones of volcanic activity around the rim of the Pacific Ocean. Where the crustal plates have moved apart, the Atlantic and Indian oceans grew, and through spreading of the ocean floor and upwelling of the mantle, submarine ridges and mountain chains were formed. The best example of these is the Mid-Atlantic Ridge. Iceland is the largest of the series of islands formed in the Mid-Atlantic Ridge and continues to be a site of volcanic activity. A whole new island (Surtsey) grew up from the ridge in 1963. From north to south across Iceland one can trace a great fissure that is actually the midline of the Mid-Atlantic Ridge. The fissure is expanding gradually all the time as North America and Europe continue to separate. (A handy comparison of the rate of movement of the plates is that they are separating at the same rate as fingernails grow.)

In the history of the earth's surface, the Atlantic and Indian oceans are quite a recent phenomena. The Atlantic seaway started to open up between Europe and Africa and the American landmasses in the late Mesozoic. In eastern Brazil there is a great set of fossil beds that was laid down along the edge of this newly opening, still-shallow Atlantic Ocean. Among them is the recently described fossil coelacanth genus *Axelrodichthys,* which dates therefore from the time just as South America was separating from Africa and just as the last flourishing of dinosaurs was occurring on land.

At the same time that the South Atlantic was being formed by the separation of Africa and South America, the Indian Ocean was being formed by the net movement of Antarctica, Australia, and India away from Africa, starting some 125 million years ago. The subplate that includes present-day Madagascar was wedged between eastern Africa and India at first, but as India moved away, Madagascar moved southward along the African coast, and it reached essentially its present position relative to Africa some 50 million years ago. Next, the Red Sea formed by separation between the Arabian plate and the African plate, perhaps starting about 30 million years ago. South of this, only some 10 million years ago, the Somali plate,

FIGURE 21 The relative positions of Africa, Australia, Antarctica, India, Madagascar, and the fragments of Asia 140 million years ago. After Besse and Courtillot.

which forms the eastern part of southern Africa and the floor of the Indian Ocean out to beyond the Seychelles, started to move away from the rest of Africa. At the western edge of the plate this movement formed the great system of faults and associated volcanic activity that we call the African Rift system.

Where were the Comores in all this? They were nowhere at all until five million years ago—or less. They are extremely modern islands formed from deep undersea volcanoes that stand right up thousands of feet from the seabed. However, they arose by a different process from the volcanic islands thrown up by mid-ocean ridges. Geologists have discovered places in the deeper structure of the earth that they call hot spots. At these points convection patterns in the fluid interior of the earth force molten magma upward into the lithosphere (the skin of solidified crust on the surface of the earth). Hot spots can burst through to produce volcanic activity anywhere in the overlying crustal plates, not just at the margins, where we see volcanism resulting from weaknesses in the active zone

FIGURE 22   The same map as Figure 21, but 85 million years ago.

FIGURE 23   The same map as 20 and 21, but 48 million years ago.

between plates. These hot spots remain more or less fixed in position.[77] When a plate moves over a hot spot, a succession of volcanoes will be formed through the crust, and the positions of the volcanoes track the movement of the plate. It was these lines of volcanic activity that caused geologists to recognize the phenomenon, and the first place that hot spots were spotted (so to speak) was the Hawaiian chain, which consists of volcanic islands formed in the center of the Pacific plate. The American geologist James Dwight Dana noticed the first evidence of this about a hundred years ago.[78] He saw that the islands increased in age from southeast (Hawaii) to northwest (Niihau) and that the younger islands were more active volcanically. Now we theorize that the islands were formed as the Pacific plate moved northwest across this major hot spot.

As it moved away from "Africa," the Somali plate passed over two major hot spots in the mantle; today we can find one (the Comore hot spot) at the western edge of the Comores and the other (the Réunion hot spot) near the southern tip of the Réunion-Mauritius-Mascarene Ridge system. These hot spots produced two chains of volcanic activity similar to the Hawaiian chain, throwing up volcanic matter from three thousand meters deep to create islands, reefs, and banks.[79]

The Comores are the newest and most westerly land thrown up by the Comore hot spot. The oldest (i.e., the first) portion of the series is thought to be the Seychelles Islands, which date from 60 to 35 million years ago. Next came the Amirantes and Farquhar islands, although there is a problem in getting absolute dates. (Dating by means of analyses of potassium and argon in the rocks depends on having samples of uncontaminated volcanic rock. Where an island is capped with coral formations and sand, the rocks that form its base may be invisible at the surface.) Next, the hot spot threw up a series of volcanoes at the northern tip of Madagascar. These are dated at about 10.4 million years and include the eroded volcanic cone that forms the wonderfully sheltered port of Diégo-Suarez. Then, as the plate moved in an arc-shaped path, Geyser and Zelée banks (where Hunt was shipwrecked) were formed. They now exist as banks of coral and sand capping the eroded tips of the volcanoes and are hard to date accurately but they must be about 8 million years old. Then the Cordelière, Castor, and Leven banks were formed (perhaps 7 million years ago).

Then came the Comores, which are geologically quite young. The oldest of the four islands is Mayotte (where the oldest rocks are dated at about 5.4 million years and the youngest at 830,000). Mohéli is dated at 2.9 million with the most recent rocks at 140,000 years. Anjouan may be about the

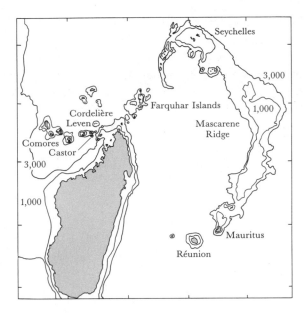

FIGURE 24   Map of the western Indian Ocean showing the islands and the thousand- and three-thousand-meter depth lines.

same age as Mohéli; the only dated rocks are 1.3 to 1.2 million years old. Grande Comore is extraordinarily young, dated at no more than about 130,000 years, and of course, the newest rocks were those formed by the last eruption of Mount Karthala, which occurred in 1977.

In parallel, the Réunion hot spot created the line of islands and banks forming the Mascarene Ridge (dated perhaps 35 million years), Mauritius (dated at 7.8 million years), and finally Réunion Island (2 million) and Rodriguez (less reliably dated at 1.5 million years). The Mascarene Ridge is related to the north-south Chagos-Laccadive Ridge off to the east. The two may have separated as a result of spreading at the Central Indian Ridge, which was formed during spreading of the sea-floor of the Indian Ocean, rather as the Mid-Atlantic Ridge was formed.

So here is a great puzzle: If the Comores did not exist until about five million years ago, where in this system of evolving ocean basin, islands, banks, and ridges did the extinct forebears of the living coelacanth live during the sixty-five million years since the end of the Cretaceous? And wherever *that* was, are they there still?

CHAPTER 4

The Comores: Catches, Observations, Early Results

Nul n'ignore plus que ce nom de
Coelacanthe designe un poisson
remarkable.

—*J. Millot and J. Anthony*

The Comoro Islands are geo-
logically diverse. Grande Comore is the youngest of the four
islands, and its largest volcano, Mount Karthala, erupted as
recently as 1977; recent lava flows abound. Anjouan, deeply

eroded and barren, is possibly next in age although there are no active volcanoes. Mohéli and Mayotte are older and less steeply mountainous as erosion has worn down the soft volcanic rocks. The shores of Grande Comore and Anjouan are rocky, and the underwater slopes very steep. The east shore of Grande Comore is particularly steep and barren. Generally on Grande Comore, at one kilometer from shore the depth is already approaching four hundred to five hundred meters. Grande Comore and Anjouan, the two islands where the coelacanths have been caught, are interesting in that they lack a typical fringing reef and lagoon. Grande Comore has no reef at all, and Anjouan has only incomplete patches of reef. Mayotte, on the other hand, has had the most complex geological history of the group and boasts a unique double reef. Everywhere in the Comores, the waters are rough and stormy, and fishing villages few and far between, scattered in the places where men can work, mostly on the western sides of the islands.

The islands are surrounded in all directions by water of around three thousand meters deep. To the east, between the Comores and Madagascar, two groups of seamounts rise with surface banks and reefs: Geyser and Zelée banks, closer to the Comores, and Cordelière, Leven and Castor banks, closer to Madagascar. To the west there is only Mozambique on the African coast. To the north-northeast are Aldabra, Assumption, Cosmoledo, and Astove islands.

The Comoro Islands have had a checkered history. The people of the Comores trace their roots principally from Malagasy, African, and Arab stocks. In the eighteenth century the Comores were important as supply stations for British and Dutch merchant ships, especially because they had freshwater. The islands were regularly sacked by pirates from Diégo-Suarez, Madagascar, and then in turn became a haven from which pirates operated, especially from the sheltered harbor of Dzaoudzi. The Comoran sultans sought protection from the European powers (some of the pirates were from the British

West Indies), and first the British and then the French started to take over the islands. The French annexed the island of Mayotte in 1843 as a counter to the British presence in Zanzibar, and they eventually developed the Comores' single most important crop, the flower of the ylang-ylang tree, which provides an essential component of perfumes such as Chanel No. 5. In 1912 the islands became a French colony, administered from the pacified port of Dzaoudzi.[80]

In 1975 the Comores declared independence, but in a 1976 referendum the island of Mayotte voted by a large majority to remain French, sending a senator and a deputy to the National Assembly in Paris. The three islands forming the République Fédérale Islamique des Comores soon fell under an extremist regime. During the resulting anarchy, landowners and religious groups were attacked, and even the elements of government (civil service, libraries) systematically purged and archives destroyed. Order was restored only by the landing of a force of French mercenaries in 1978. Because the Comores occupy a key strategic position at the north end of the Mozambique Channel, both Western and Middle Eastern powers, as well as South Africa, have a strong interest in their remaining pro-Western.

Early in 1990 the Comores were again in turmoil. When the restored president, Ahmed Abdallah, was assassinated in November 1989, the mercenaries were evicted by French forces.

The Comores, a desperately poor region, labor under one of the highest population densities in Africa. The economy has always depended on cash crops, particularly the distilled oil of the ylang-ylang. But the need for food crops has eroded that economic base; the food crops recently planted instead of the ylang-ylang cannot produce enough to feed the population, and there is no money to buy imported food. Deforestation and soil erosion are now widespread problems and getting worse. As we will see in the last chapter, the poverty of the islands makes the status of the coelacanth extremely precarious. In this context of deprivation the fish is a potential eco-

nomic resource. A local fisherman can sell a coelacanth either
to the government or through a black market for more than he
could normally earn in a year. Before Smith and later Millot
circulated their posters, however, the fish had no commercial
value, its flesh being too oily to eat. The modest amount of
fishing that is possible from Comoran fishing villages was traditionally concentrated on a very few species, and even then
the fishing off the Comores is generally poor.

A lot of different scientists have been interested in coelacanths since 1952, and many have traveled to the Comores to
try to study them at first hand. In the years between 1952 and
1986 a succession of expeditions fished for them, using modern methods and gear, trawls or deep lines, submersibles, shallow diving, and deep, remote photography, but none was successful. It is a remarkable fact, however, that in the fifty years
since the first living coelacanth was found, only the very first
specimen was caught by modern methods; even so, being
taken in a trawl, the fish was in a sense bagged by accident.
Otherwise, "technology" has failed. All subsequent specimens
have been caught by the native fishermen of the Comores
using their traditional methods.

Fishing off the islands is not easy. Nowadays many fishermen operate fiber glass boats with outboards, but the basic
fishing craft for hundreds of years has been, and still is, the
dugout canoe, hollowed from a single log of mango or kapok
and steadied with one or two outriggers. It is a good one- or
two-man vessel for work both within the reef and, to a lesser
extent, outside. In the Comores one sees a range of canoes
from smallish one-man types to heavier ones that are used to
carry small cargoes.

On Grande Comore and Anjouan, where there is no sheltered sunlit lagoon that might be a ready source of food fish,
the fishermen must work in more exposed and difficult conditions, where the chance of getting a decent catch is indeed
poor. Here most fishermen have traditionally used a very simple tackle, basically a long hand line. The fishermen usually do
not fish in more than one hundred to two hundred meters, and

they usually need not venture more than a kilometer from shore to do so. They use a single line with a single hook, baited with a chunk of some fish like tuna or flying fish. A favored bait for the coelacanth is said to be the roudi *(Promethichthys promethus)*, which comes from the same depths as *Latimeria*. The hook is a commercial hook on a thick nylon leader, and near the end of the line itself is a short length of string, onto which a chunk of larva rock is knotted with a slipknot. The line is payed out, and when the right depth has been reached, the fishermen gives it a jerk to release the rock weight. The baited hook then hangs free just above the bottom, and sometimes a coelacanth will take it and be brought up.

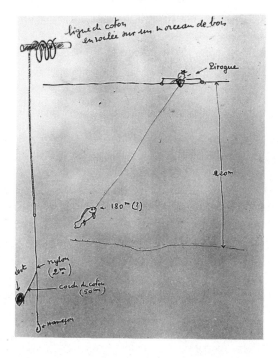

FIGURE 25   Sketch of the traditional longline technique, sent to me by a Comoran official in 1966.

The coelacanths that have been caught so far range up to
about 180 centimeters and 75 kilograms. So large a fish must
be played very carefully by the man in his pirogue. It may take
more than a hour to bring it up. He cannot afford to lose the
gear because a line of 300 meters might cost the equivalent of
six months' earnings. Any big fish being brought up is likely to
be valuable. If it is an oilfish *(Ruvettus pretiosus),* it will be valuable at the market. If it is a coelacanth, however, there is the
big reward, the equivalent of winning the lottery.

Once the fish is up to the canoe, a prudent fisherman will
(unfortunately) usually immobilize it with a blow to the head.
A big fish might even be too big and ferocious to be brought
on board, so a big hook will be stuck through the lower jaws,
and it must be towed back to shore at the risk of being
snatched by a shark. This may take another hour or more.
Then the alarm is given at the village, and the prize catch
inspected. Occasionally, and to the very lucky, a catch will happen while a scientist is there. He or she has alerted the fishermen all along the coast that a coelacanth is needed, alive if
possible. Someone runs to the hotel. The scientist comes
down to the water's edge, and there in the shallows is the
fabled fish.

This happy scenario—of a fish being brought to shore still
alive—has been played out at least five times. As already re-
counted, the first time was in 1954, quite early in the French
research effort. It happened again in 1972 during a joint Royal
Society/National Academy of Sciences/National Geographic/
Muséum National d'Histoire Naturelle expedition that will be
described in the following chapters.

In July 1966 *Life* magazine published an extraordinary story
and accompanying pictures, purporting to have images of a
live coelacanth "in its dim and aerie habitat" off the Comores.
French photojournalist Jacques Stevens, the story said, was
diving at night when "out of the spooky darkness at about 130
feet below the surface, here came a coelacanth right at him."
The fish was "mucus-covered," and "its huge phosphorescent
eyes glared at me. . . ." Steven's movie camera jammed after he

had managed to get only a few feet of film, and the flash of his still camera "confused and bothered it temporarily. Then the coelacanth swam off, disappearing into the depths again."

The pictures Stevens published are indeed impressive and certainly are the first of a live coelacanth. However, as soon as we saw the two photographs, scientists everywhere realized that the story had been unnecessarily embellished; a lot of it could not be true. The fish, shown against a background coral that could be found only in much more shallow water, was lit not only by the photographer's flash but also apparently by sunlight. The snout of the fish showed vertical lines where the pigment was missing, which is typical of the effect produced by the rubbing of a fishing line. The operculum was red with blood, suggesting stress or asphyxiation. The eye was cloudy, suggesting that the fish had been subject to a rapid depth change. The fact that the fish was "mucus-covered" was by itself a giveaway. Clearly this fish had been caught on a line in the usual way and brought to shore after some effort. It was probably alive when released and photographed, but it had been close to death.

Stevens did make some observations of the fish swimming. The tail was "not used for propulsion but acted as a keel. It moved chiefly with its second dorsal and anal fins, using the pectorals as stabilizers and for turning movements."

In 1979 another line-caught fish was photographed. The British Broadcasting Corporation had sent a camera crew to the Comores seeking footage of a live coelacanth for its ambitious "Life on Earth" project. Although the crew failed to get live pictures with a towed deep TV camera, photographer Peter Scoones got some very nice pictures of a live coelacanth that were published in magazines around the world. He had been on Grande Comore for the BBC when local fishermen brought a specimen in and tied it up under the shade of some pirogues. Scoones released it and tried to revive it by passing water over the gills. He managed to get a few nice pictures of it, even though the fish was in an exhausted state.[81]

In the years between 1953 and 1963, when the French were

actively collecting the coelacanth, twenty-three specimens were taken, but only on the one occasion in 1954 did they observe a live one in a systematic fashion. Several more coelacanths may have been alive when brought to shore; but the scientists were now in Paris, and the fish was preserved immediately. Obviously it was too costly a matter to station a scientist permanently on each island on the rare chance that a specimen might be caught. And distances from other places where scientists normally work were too great for someone to fly in when called. So each specimen was preserved in formaldehyde—the sort of the embalming fluid that ichthyologists use for all specimens—and sent to Paris, to the Laboratoire d'Anatomie Comparée, of the Muséum National d'Histoire Naturelle.

In Paris the research effort set into place to describe the anatomy of the coelacanth began to produce results. Over the years dozens of research papers and three volumes of a profusely illustrated monograph, *L'Anatomie de Latimeria chalumnae,* have been published.[82] The skeleton, musculature, nervous system, sense organs, reproductive organs, and digestive tract have been described in detail. In recent years the work has been continued by Dr. Daniel Robineau and others. It has been a mammoth undertaking and perhaps one of the last great exercises in formal descriptive anatomy.

While the scientific community has appreciated the French anatomical work, there was much more to learn. For example, we wanted to know; What were all these muscles for? How did coelacanths swim? Where exactly do they live? They have the curious rostral organ, but what do they use it for? How do they reproduce? How do they feed? What do they feed on? At the same time no one else was yet permitted to study *Latimeria chalumnae.* So other sorts of analysis had to be attempted, by working secondhand, as it were, from interpretations based on the anatomy.

Evidence concerning the living biology of coelacanths remained frustratingly sparse until 1966. It was as if the superb specimens in Paris were somehow only more elaborate fossils.

The information that could be gathered from them was at the same time extensive and limited.

Pieces of the puzzle slowly emerged. From the specimens that the French dissected, a listing of stomach contents was assembled. Coelacanths had fishes and squid beaks in their stomachs. The fishes came from a range of habitats. Some were fishes of sandy bottoms of depths around two hundred meters or so. Some were mid-water fishes, members of the Myctophidae—modest-sized fishes (about four to six inches in length), particularly interesting because of their very special behavior.[83] Such fishes undertake daily vertical migrations through the water column—going deep during the day and rising toward the surface at night, and back again. They track a narrow range in level of illumination, and they accompany a vast cloud of invertebrates that perform the same diurnal migration, feeding on them and in turn being fed upon by bigger fishes. This joint migration is so immense in some areas of the ocean that it readily shows up on the sonar detection devices of research ships. The phenomenon was discovered during the Second World War by U.S. Navy scientists experimenting with sonar who called it the deep scattering layer, and its diurnal movement was recorded long before anyone realized that it consisted of clouds of organisms—an organic Milky Way of food and predators in the ocean.

Where in this migration do coelacanths feed upon mid-water fishes? If *Latimeria* regularly feeds on these fishes, does it also make a diurnal migration? Or is it more sedentary, waiting to take food as it goes by?

The French group's conclusion concerning the coelacanth's peculiar intracranial joint was a surprise to many.[84] It concluded that the joint was immobile; no movement was possible there, and it must be a relic of something else, rather like the human appendix.*However, this conclusion did not match

---

*In fact, from comparative embryological studies it is known that in all vertebrates the site of the intracranial joint must be the place where two major components forming the basal regions of the front and back parts of the neu-

with one highly interesting result of the descriptive anatomical study of *Latimeria*. This was the discovery, first made by Smith and then confirmed and amplified by the French group, of an entirely new paired muscle, extending across the joint, on either side of the ventral midline of the skull, and connecting the two hinged sections. Not only is there a joint, but there is a muscle presumably functioning to rotate the anterior region of the braincase downward and backward.

A fascinating fact about the vertebral column in *Latimeria,* as in all lobe-fins, is that it consists only of some very lightly ossified bony rings around a stiff but elastic, fluid-filled notochord. Again this seemed a throwback to an embryological condition, for the notochord is a very primitive structure both in vertebrate phylogeny and in vertebrate development. The sort of bony backbone that we have is a later development. In early vertebrates the notochord was the principal structure in the "backbone" region and was such a dominant feature that it gives the whole taxonomic group to which we and the fishes belong its Latin name: the Chordata. In most advanced vertebrates the notochord forms early in development, but then its growth stops and the massive bones of the vertebral column grow around it.*

The notochord in *Latimeria* is not only large but actually hollow and is filled with a liquid, as was noticed by Courtenay-Latimer and the taxidermist when they cut open the first speci-

rocranium (braincase) fuse together in early development. So one theory was that the persistent joint must be an embryological oddity caused by something that prevented that fusion.

*These bones normally both replace the notochord as the principal structural element of the "spine" and grow up as arches around the dorsal nerve chord (or spinal column), forming a complex structure, the vertebral column or backbone. In the backbone of reptiles, birds, and mammals (including, of course, humans) the notochord is all but completely replaced by the backbone in adults, but a remnant of it persists between the bony vertebrae as the notorious intervertebral disks, a sort of shock-absorbing washer between the vertebrae. When a disk is injured, atrophies, or slips out of place, it may cause pressure on the nerves that pass out from the spinal column between the vertebrae to serve the limbs and other organs; the result is terrible pain.

men. This feature is probably unique to the coelacanth group. The nature of the coelacanth notochord must give special properties to the way it swims, although its precise significance is still not known. The notochord is the functioning backbone in the adult coelacanth. It is tough and elastic and possibly acts as a sort of spring. As the body is flexed from side to side, energy must be stored in the spring and then released.[85] Interestingly, the notochord is also the major part of the backbone in lungfishes (although it is not hollow, at least in the living forms), and this may mean that all these fishes arose before a complex bony backbone had been evolved by vertebrates. The *bony* backbone then later developed twice in parallel—once in higher ray-finned fishes and once in lobe-fins/tetrapods.

In coelacanths the notochord also passes all the way forward underneath the posterior half of the braincase and fixes to the back end of the front half. It must therefore also participate in some way in the operation of the intracranial joint, and this attachment of the notochord could be taken as lending credence to the notion that the joint is immobile. One theory developed that perhaps the joint served only as some sort of shock absorber against the forces developed by the jaws in biting. On the other hand, the presence of the large specialized subcranial muscles argues against all these "passive" theories. The trouble was, one had no way of checking just how immobile the joint actually was. Specimens fixed in formaldehyde soon become very hard, especially their ligaments and connective tissue. In all the preserved specimens the joint would indeed be immobile, as the French concluded. No one had checked it in a fresh specimen.

From the anatomy of the fins, scientists began to speculate about the mode of swimming. The massive tail, flattened from side to side, was obviously very powerful but did not seem very flexible, and it is certainly not the sort of tail that one finds in fish that show continuous cruising swimming at moderate or high speeds. When the French observed the live specimen caught in 1954, they saw that it swam slowly "by curious rotating movements of its pectoral fins, while the second dorsal and anal, likewise very mobile, served together with the tail as a rudder."[86] And what about the stubby muscular nature of the paired fins? From comparison with the living lungfishes and from analogy with tetrapod vertebrates such as Amphibia and from other fishes with various versions of "legs," like the mud-skipper and the walking catfishes (both very advanced ray-finned fishes far removed from lobe-finned relationship), it originally seemed that the limbs might be used to push off against a solid substrate. They are not very strong and could not deliver a powerful thrust, but of course, a fish in water is essentially weightless. All this seemed to fit with the notion that the fish lives on or near the steep-sided submarine slopes of the island bases.

Reproduction was an interesting problem. Dissections of the reproductive tracts of both males and females had been made and published, and the evidence they offered was quite conflicting with respect to the question of whether the coel-acanths are live-bearers, as the fossil evidence had suggested to Professor Watson in 1926. The female tract was interesting in lacking any kind of shell glands. Therefore, if eggs were laid into the open, they would have no shell but would be more or less defenseless.[87] On the other hand, there seemed no possi-bility that the females were live-bearing because the males have no intromittent organs, no equivalent of a penis, by which to place sperm into the female reproductive tract. The males are characterized by a rosette of caruncles, spiny scales around where the reproductive, urinary, and alimentary tracts open to the exterior. But that seemed unlikely to constitute an intro-mittent device. It looked as though, as is the case with the vast majority of all fishes, the eggs and sperms were shed together in the water. Fertilization must therefore occur in the water, and the eggs must grow up more or less unprotected. In that case some kind of parental care would be indicated. For exam-ple, the lungfishes, cousins of the coelacanths, are known to lay their eggs in nests guarded by the adults. Perhaps this might also give a special set of additional functions for the flexible and elaborate paired fins. This set of conclusions was, if not confirmed, at least not refuted by the discovery inside the eighteenth French specimen (a 180-centimeter fish, caught on January 1, 1960) of a large number of eggs, mostly small and strangely colored, but a few of which were about the size of a hen's egg (seven centimeters in diameter).[88]

By 1972 a significant number of coelacanths had been caught and preserved—sixty-eight to be exact (but two were lost, and for two the locality was unknown)—and the French team had been very careful to keep records about each capture. This was enough for it to begin assembling some interesting "vital statistics" about the population of fishes on the Comores.[89]

Although there are four islands in the Comores group, *Latimeria* had been caught off only two of them—Grande Comore and Anjouan. Most of the specimens (forty-four out of sixty-four) were taken off Grande Comore, and all but two were landed at villages on its western shores. Although specimens had been taken throughout the year, most catches had occurred in the months of December through March, and all were made at night. The general reported depth range was from a minimum of eighty meters to a maximum of six hundred meters. In fact, the French had also made a concerted effort to get the local fishermen to fish deeper than their usual hundred meters or so. When they did, they still caught coelacanths—all the way down to six hundred meters. But the early catch rate was still relatively low—some two to three per year on average, with a maximum of eight in 1965. At this point, for reasons that will be explained, political events gave a new impetus to the local fishing efforts.

The fact that the fish were caught on only two of the islands seemed to indicate a very limited distribution in space, and the fact that the catches were also seasonal in nature also seemed to indicate that the fishes undertook some kind of geographic migration. That they are caught at night possibly suggested that they live deeper during the day and come nearer the surface at night to feed. This seemed to be confirmed when fishermen were encouraged to go deeper. Six hundred meters is a very long line to work by hand from an outrigger canoe.

By the mid-1960s many zoologists thought that they had reached a limit on what could be learned from dissections of preserved fishes, and we dreamed instead of ways of getting closer to the living fish. Captain Jacques Cousteau had explored the Comores with his submersible as early as 1954 but had seen no coelacanths. In 1955 a group from the Steinhart Aquarium in San Francisco tried to organize an expedition to capture a specimen and keep it alive, but the French denied it permission. Then a slight opening presented itself, and here the story becomes directly the story of this author, who was one of those dreamers.

In 1965 I was a twenty-seven-year-old postdoctoral research fellow at University College of London University. I had completed my Ph.D. at Harvard under the famous zoologist and paleontologist Alfred Sherwood Romer (one of those responsible, as it happens, for the shock absorber explanation of the intracranial joint) and was conducting research on fossil fishes in London. I had just completed a study of the intracranial joint in the group of fossil lobe-fins called osteolepiform rhipidistians, the fishes that are most commonly thought to be the ancestors of tetrapods (Chapter 3). I was trying to understand two things: what the biomechanical function and significance of the joint might have been and how the joint region had become transformed in the transition between lobe-finned fishes and first land amphibians as the fishes involved changed their feeding and respiratory mechanisms.[90] As a result of all this work, I had become interested in the joint in *Latimeria*, the only available living model. I was, to put it mildly, challenged (irritated might be closer) with the statement, now firmly established in the zoological literature, that the joint in coelacanths was immobile. It seemed to me intuitively obvious that such a complex structure of bones, joints, notochord, and muscles must have had a specific function and that there must have been significant intracranial movement. I made mechanical models of the skulls of the fossils and of *Latimeria* and worked out a hypothesis of how they might have moved and what the joint might have been for (see more details of this in Chapter 6).

As I readied that study for publication, I received an offer to go to Yale University as an assistant professor of biology and curator in the Peabody Museum of Natural History, and my wife and I moved to New Haven, Connecticut, in the summer of 1965. In late January 1966 Professor Elwyn Simons, then a paleontologist in the department of geology, came to me with a simply amazing document, a letter from the Ministère de la Production et des Industries Agricoles of the Comoran government asking if we would be interested in purchasing a specimen of *Latimeria.* Simons had held the letter for a little while before he thought to ask me about it. It was now a few weeks old, and identical letters had obviously been sent to every major zoological institution in the world. Nothing daunted, I asked the director of the Peabody Museum for approval to try for a specimen. The price was so absurdly cheap (four hundred dollars plus shipping) that I was quite prepared, if necessary, to shame the university by paying for it myself if funds could not be found. Shades of J. L. B. Smith!

The background to all this was that the Comoran government was steadily moving toward independence from French colonial rule and among other things wanted to capitalize on one of its major assets as a way of raising some foreign currency. The French group in Paris had decided that it had sufficient specimens for its anatomical studies which were in any case approaching completion. So the Comoran authorities offered coelacanths for sale, if and when they became available. They had managed to catch eight in 1965 and expected more in 1966. The French monopoly on the fish was apparently broken, and restrictions about using specimens for research instead of simply for exhibit were lifted. It is for this reason that the numbers of coelacanths caught off the Comores then reached an all-time high. Not surprisingly, the Comorans had an overwhelming response to their circular letter. We were very late in this rush.

Luckily Dr. Alfred W. Crompton, then director of the Peabody, a South African paleontologist who needed no persuading of the coelacanth's importance, generously encouraged the effort to get a specimen. For my part, I had no hesitation about what I wanted. I wanted a coelacanth, to be sure, but I wanted a fresh one. Another formalin-preserved specimen would be much less use. Obviously the only way to get a fresh fish was to have one shipped over frozen. Naively I fired off a response to the ministry: Yes, please send us a coelacanth, but frozen!

Our request reached the ministry, and by rights, it should have been right at the bottom of the pile. We were late, and at the usual catch rate of fewer than five per year, it should have been many years before we got our specimen. But although we did not know it, ours was right at the top of the list for *frozen* specimens. It was the only one.

In mid-March I got a letter from the Comores saying that things looked hopeful and one from the U.S. Embassy in Madagascar saying that transport of a frozen specimen would be impossible. Then, on April 5, 1966 (Good Friday), I got a message at home to call the U.S. State Department in Washington. I returned the call, in some trepidation that perhaps something had gone wrong with my alien registration papers, about which dire warnings were broadcast each January in those days. Instead, I reached a very helpful assistant somebody or other who told me of a cable received from the consulate in Marseilles. It seems that the SS *Pierre Loti* had reached Marseilles, where it was due for a refit. The freezers were about to be turned off. Aboard had been a mysterious large box picked up from the Comores. Except that it was addressed to the Peabody Museum, Yale University, the box had no further shipping instructions. Did I know anything about it? Well, no . . . but it could only be a coelacanth. D. V. Anderson, the U.S. consul general in Marseilles, had alertly taken charge because with the long Easter weekend coming and no refrigeration, it would otherwise surely have spoiled. A consular vehicle had taken the box to a commercial frozen food warehouse for storage, and by the time I called, arrangements had already been made for its shipment to New York. (The local newspaper *La Provençal* heard about our coelacanth and lamented the fact that this specimen was not going to the Marseilles Museum of Natural History.)[91]

A few days later a letter arrived from the Comoran government telling me that a specimen had been caught on March 14, immediately frozen, and shipped off on a conveniently passing fruit boat on March 19. Passage was paid as far as Marseilles; the rest was up to us. This letter had been sent out on the *following* boat, however.

With the magnificent arrangements made by the consulate in Marseilles, the small thirty-nine-inch specimen was soon safely at Yale, where a minor circus developed. Many people expressed great interest, and my wife, Linda, pointed out that even more would like actually to see such a marvel. So for a few days it lay in state on a small bier (actually it was in a borrowed ice-cream freezer case with a glass window), while most of Connecticut filed past. It was a miniature version of what J. L. B. Smith must have enjoyed/endured. The Great Hall of Yale's Peabody Museum came to resemble nothing less than Lenin's Tomb. Reporters flocked for interviews. Meanwhile, my colleagues and I planned. We could thaw the fish only once, when every possible tissue sample must be taken and prepared. We got in touch with everyone we could think of, taking as our model the way geologists had recently planned the distribution and study of the first rock samples brought back from the moon. Then, having prepared as well as we could, we thawed it out.

Would there be an appalling smell, the result of the specimen's having sat unattended for days on some tropical dockside? No, it was fresh as a striped bass straight from Long Island Sound (and perhaps a little less dosed with pesticides). For the first time we could study its blood, its biochemistry, its physiology. This one specimen, the "Yale Specimen," opened the door to all further studies of the coelacanth.

A new phase of research on *Latimeria* had been launched. As soon as the results from our "fresh frozen" specimen started to come in, more and more zoologists around the world became energized to ask and answer their own pet questions. Other institutions requested and obtained frozen specimens, and plans were hatched yet again for expeditions to the Comores to secure truly fresh specimens—perhaps even a live one.

In 1968 plans were made for a joint expedition to the Comores by the Royal Society of London and the Muséum Nationale d'Histoire Naturelle in Paris. The long negotiations have been delightfully described in a book by Dr. Jean Anthony, where we learn the surprising fact that despite all the specimens held in Paris, new material was needed even for gross anatomical studies. It seems that many, if not most, of the specimens were poorly preserved: *"Depuis une quinzaine d'années je déplore de recevoir des spécimens défectuex . . . ce matériel mediocre."*[92]

Plans were completed for an expedition in 1969, except for a crucial matter: At the last minute the Comoran government refused to grant permission to fish for *Latimeria*. So, without the French, the expedition was recast to answer one of the major problems: geographical distribution. The strategy of the Western Indian Ocean Deep Slope Fishing Expedition of 1969 was to fish intensively along the key islands and banks of the western Indian Ocean immediately to the north and west of the Comores, to see where else *Latimeria* lives. Dr. G. R. Forster from Plymouth, England, led the group, and I took part as a U.S. representative for the National Academy of Sciences. We were also ready, if we caught a specimen, to observe it and then prepare tissue samples for study. The expedition worked southeast from Aldabra Island to the two sets of banks (Zelée and Geyser; Cordelière, Castor, and Leven) east of the Comores.

The method we used was basically a mechanization of the old ways of Comoran fishermen. We laid down dozens of deep lines each night, each with baited hooks suspended at a range of distances from the bottom. Ideally we could fish at a series of intervals from the bottom. The fish faunas of these islands turned out to be impressively broad in both numbers of species and the size of our catches. We caught a huge range of every imaginable—and some almost unimaginable—fishes, but not one coelacanth. We caught the oilfish *Ruvettus pretiosus;* we caught *Pseudotriakis microdon,* a rare shark; we caught a new species of *Squalus* and a new arthropod. We caught hundreds of *Etelis marshi,* a kind of red snapper that the reference works said had only infrequently been caught in that area. *Etelis marshi* was so abundant, in fact, that we were able to recommend the founding of a new commercial fishery using longlines.[93]

Some of our fishing was done with hand lines from a small dinghy. We did this close inshore and at shallower depths than the majority of the lines. Floating on the glassy, still Indian Ocean, we were visited by several sharks—oceanic white-tips of the species *Carcharodon albimarginatus,* known as a man-killer. One shark in particular I remember because when its head had gone past the bow of our fourteen-foot boat, its tail still had not arrived at the stern. We were fishing off Geyser Bank and although we did not know it, not far from where Hunt and his crew had died. Despite all the abundance of fish we caught, however, there was no coelacanth. And negative evidence being what it is, we still do not know whether the coelacanth might live off those banks and islands. No one has tried there since.

In 1972 the same group organized another expedition, this time with full participation of the Muséum National d'Histoire Naturelle as well as with financial support from the Royal Society, the museum, and the National Geographic Society. This time we obtained permission to fish, both by using the deepline method developed for the 1969 expedition and by encouraging the local fishermen. The hope was to be right on the spot to deal with the catch, and once again there were detailed protocols for the preparation of tissue samples.[94]

This expedition was a fabulous success. All in all, it stayed out for three months and took two extremely important specimens. The first members of the expedition to arrive were the French and British; funding for the American group had been delayed. In fact, the group had hardly arrived when the first specimen was caught, almost taking it by surprise. On January 5, while the expedition was still setting up its base at Moroni on Grande Comore, word came that a local fisherman had taken a specimen off Anjouan. It was still early morning, and hiring a plane to fly over to Anjouan proved tricky. However, after many complications Professor Anthony took charge of this new specimen, and the expedition was already a success.

Like the first of the Comore specimens, the new fish had been taken near Domoni, at about one in the morning, 2 kilometers from shore and in a depth of about 400 meters. It was a big fish, some 163 centimeters long and weighing 78 kilograms. When the scientists arrived, the proud fishermen told them that it had stayed alive quite a long time. Regrettably it was now dead, but there it was—a magnificent female specimen. Upon dissection this fish provided a great deal of the special fresh tissues the expedition wanted and one new discovery. While one of the early specimens taken by the French had eggs described as being like chicken's eggs, this one had nineteen huge eggs, the size of grapefruit, about 8.5 to 9 centimeters in diameter and each weighing some 300 to 350 grams. These were probably the largest eggs of any fish in the world, matched in size only by those of a few sharks.[95]

FIGURE 26 Two of the "grapefruit-sized" eggs from the January 1972 specimen. COURTESY OF LABORATOIRE D'ANATOMIE COMPARÉE DU MUSÉUM NATIONAL D'HISTOIRE NATURELLE, PARIS, DR. DANIEL ROBINEAU

The expedition then waited with high hopes—and waited. The scientists tried the experimental fishing program without success. Not only did they fail to catch a coelacanth, but they caught few fish of any sort. The American contingent joined the expedition in February, by which time some of the original group, including Anthony, had left for home. To my everlasting disappointment, because of the delays, the expedition now overlapped with the Yale University spring semester, and the (as I thought mindlessly uncooperative) university administration refused me permission to make the trip. The British fishing team eventually returned home, and the remainder of the expedition now had to rely totally on the local fishermen. By mid-March the expedition was continuing to disband. Only three members remained: Drs. Daniel Robineau (Paris), Bob Griffith (from my laboratory at Yale University), and Adam Locket (Institute of Ophthalmology, London). On March 21 Robineau left for Paris, leaving only Griffith and Locket to wind up the expedition. Then, on March 22, Bob Griffith sent me the following cable: "Live coelacanth caught this am movies dissection OK stopover in Paris home around the First. Bob."

I was in a better position than J. L. B. Smith had been when he got his letter from Courtenay-Latimer or his cable from Hunt. At least I knew that the fish was in safe hands. But I waited anxiously for details. Alive! Movies! It was even more exciting than the day we had driven the frozen specimen home from the New York docks to New Haven or the day we had thawed and dissected it.

The new fish was taken from the village of Iconi, Grande Comore, not far from the expedition's base, and Griffith and Locket raced to the village as soon as they got the word. The fish had been taken at about 2:00 A.M., 600 meters from shore, at a depth of 165 meters. The bait was a chunk of tuna. Iconi, like Domoni, is one of the villages from which a considerable number of coelacanths have been taken over the years, and this fisherman, Madi Yousouf Kaar, had previously caught two other coelacanths himself. So he knew exactly what he had. He

towed it carefully to shore. The village chief was roused from
sleep and set off by taxi to summon the scientists at the hotel.
All this took several hours, but it was still dark when Griffith
and Locket reached the fish, which had been placed by the
villagers into a primitive cage (made especially in case a live
specimen should be caught). Griffith and Locket observed the
fish with flashlights as dawn broke. Then it was transferred to a
fiber glass tank that the expedition had prepared for this purpose, and it was filmed while slowly swimming. The coelacanth
swam gently in the tank for some four hours before it became
moribund and was sacrificed.[96]

After the stunning success of 1972 groups from many institutions—the California Academy of Sciences, the Vancouver
Public Aquarium, and the New York Aquarium, for example—
made expeditions to the Comores and obtained frozen and
preserved specimens, but the greatest new development has
been that of Dr. Hans Fricke. In our story up to now fortune
has favored the individual rather than the conformist, and Dr.
Fricke certainly is a worthy successor to Smith. I suppose many
of us dreamed at one time or another of chartering a major

FIGURE 27   Drawing made from a photograph of the second specimen caught in 1972.

research submersible, such as Woods Hole Oceanographic's
Alvin, for an underwater search for *Latimeria.* But the glint in
the eye was quickly extinguished by the logistics, the enormous cost, and the virtual impossibility of gaining governmental funding. Dr. Fricke is a physiologist, an ecologist, and a

filmmaker at the Max Planck Institute in Seewiesen, West Germany. His approach to the problem was simple and straightforward. He got some private funding, designed and built his own two-man submersible, chartered a large yacht as the support ship, and set off for the Comores. Of course, an enormous amount of planning and work went into the venture, but its sheer audacity is breathtaking. If he had failed, he would have been just another in the list of dreamers. He succeeded brilliantly.

In December 1986 and January 1987, then again in April and May 1987, Fricke took his submersible, the *Geo* (named for the magazine that supported him), to the Comores. Armed with the previous catch records, he dived at the places and depths where the Comoran fishermen have their successes. On January 17, 1987, Fricke became the first man to see a coelacanth swimming freely in its native environment.[94] All in all, Fricke and his group succeeded in finding a total of six specimens. They saw them swimming but not feeding. In these first observations Fricke did not see any interactions between coelacanths or between coelacanths and other fishes. He observed some new behavior, however, and he took some magnificent films of them, the first really good films of live coelacanths, revealing yet more of the biology of the fish. They cruised very slowly or kept station in the currents, hovering just above the bottom. Large, bright blue, with pale spots, they didn't take much notice of the submersible or, indeed, of one another. His films show them as quiet and slow while their fins, spread wide and fanning against the water, almost seem like delicate slow-motion wings. He also recorded the sorts of environments in which they occurred. As we shall see, these results were invaluable in sorting out the confusing complex of inference and indirect data about the behavior and ecology of *Latimeria* with which we all had been working to this point. Fricke has continued the work with a new submersible, *Jago,* with even more splendid results. He has now photographed more than forty different coelacanths and will soon be able to publish observations on behavior of the fish, including social interactions. So new chapters continue to follow in the difficult but endlessly fascinating task of learning more and more about *Latimeria chalumnae.*

PART TWO

ANSWERS
AND
QUESTIONS

CHAPTER 5

Bringing the Pieces Together

They will have the satisfaction,
denied to most workers on fossils, of
having their researches fully
corroborated.

—*E. I. White*

$\mathbf{I}$n any good detective story
every fact that is established, each piece of the puzzle put into
place immediately closes off certain possibilities and raises new
questions. The skilled author then weaves a simple ironclad

solution from the tangle of evidence and theorizing. In comparison with a neatly solved mystery novel, this is only half a book. In 50 years of study of *Latimeria chalumnae* and more than 150 years of studying its fossil relatives, we have reached only that stage where the first answers suggest important new questions. Given the difficulties and the particular history of the venture, however, we may claim to have made decent progress even to have got to this point.

Just to watch the films that Dr. Fricke made of live coelacanths for a few moments confirms the impressions built up over fifty years of study of this beast. It is a large, sluggish, slow-moving animal. It is weird-looking, but then any animal looks odd if you study it hard enough, just as any word looks unreal if you stare at the page; try *yacht,* for example, or *above.* Above all, it is a *fish.* It is not a fish trying to be an amphibian or an amphibian trying to be a fish. It is a perfectly good fish and therefore must be discussed in the same terms as any other fish.

We have already noted the results of a lot of direct observation of coelacanth biology. But in a work like this we also have to resort to hypothesis and inference, particularly involving argument by comparison. For example, in the next chapter, I will argue from the evidence of the light absorbency peaks of the cells in the retina of a dead fish's eye that it lives in a particular light regime and therefore at particular depths. The argument depends on the fact that other fishes with the same retinal cell properties live under those specific conditions, and no others. It is not unlike the use of laboratory animals for testing drugs. If a drug causes cancer in a mouse, it has a certain likelihood of causing cancer in a human. If it causes cancer in a monkey or a chimpanzee, because these organisms are more closely related to humans than mice are, we can argue that the chances are stronger that the drug will be dangerous to humans. Testing on a cockroach might tell us nothing. In view of the importance of such an argument by comparison and analogy, it is worth examining the method a little further before we come back to the biology of *Latimeria chalumnae.*

Science, biological science most of all, is usually not the sort of hard-edged exercise that is so glamorous to the popular press. The scientist does not carefully amass perfectly lucid arguments based on crystal-clear facts as if he or she were plucking diamonds from the bed of a stream. Instead, the facts have to be teased out gradually. Often what we need to know is only partially revealed, and we continue to add to it slowly. What we think is a *fact* may change year by year. The personality of the investigator is important in this sort of work because he or she controls not only what is studied and how it is done but, in the end, what is actually revealed.

Biology is the quintessential comparative science. Because of the evolutionary thread that links all organisms in one great network of genetic relationships, there is an underlying orderliness to the vast numbers of living and extinct organisms. There is a set of historical (actually genealogical) patterns in the record of biological diversity. In order to understand how evolution (change over time) has worked, we first have to decipher these patterns. Only then can we discover the underlying mechanisms that produced the change. In order to understand the patterns, we have to compare. If we know that our data must fit into distinct patterns rather than be random, we can use comparative judgments to fill gaps in our knowledge. Simply by looking at a photograph of a particular species of bird, or a dead specimen in a museum, or even a fossil, we can tell a great deal about what sort of bird it was because of what we know about similar birds. With a beak of one shape it must be a seed eater; with another shape it must catch insects, or, with a third, fish. With legs that long it must be a wader and could not nest in trees, and so on. Comparative biology is a matter of constantly building up analyses of patterns.

In a case like that of the living coelacanth, the comparative method works in two directions. We already know a lot about the biology of *Latimeria* from direct observation. Until we can learn more, we must use comparative data from other fishes in order to make intelligent hypotheses about those parts of the biology of *Latimeria* that we have not yet observed firsthand. Then we can use this mixture of fact and hypothesis to provide a point of comparison for understanding other fishes, particularly the fossil lobe-finned fishes. Perhaps we can even illuminate aspects of the old question of how tetrapods arose from fishes, way back in the Devonian. Paleontologists always have to work like this; that is why living fossils are so important to them. The trick is not to confuse the two aspects of the exercise, not to transform an hypothesis on the one hand into a fact on the other.

Consider, for example, this sort of choice between hypotheses. *Latimeria* has an oil-filled swim bladder; fossil coelacanths seem to have a similar sort of structure that we can identify also as a swim bladder. In that case one could conclude that the swim bladders of the fossil species must also have been oil-filled, and therefore, they must have had a similar swimming behavior and ecology. Lacking direct evidence on the fossil forms, however, how is one to choose between this argument and the following? The other fossil lobe-finned fishes (lungfishes and rhipidistians) included the ancestors of tetrapods and therefore must have had air-breathing lungs and have lived in shallow waters. If this is the primitive condition, the early coelacanths must also have had air-breathing lungs. Further, most fossil coelacanths lived in shallow waters like the lungfishes and rhipidistians, whereas *Latimeria,* which has an oil-filled swim bladder, lives in deeper waters. Therefore, the oil-filled bladder is a secondary specialization, a feature probably restricted to *Latimeria* and to any other fossil coelacanths living in deeper (that is, not shallow or surface) waters. On balance, the evidence and the argumentation support the second hypothesis.

When scientists try to build up a picture of an organism's biological function, living or fossil, on the basis of incomplete direct information, much can be accomplished by arguments concerning physical constraints (the properties of materials and the mechanics of how things work). Suppose, for example, you had never seen a living horse. A careful examination of the skeleton of a horse would convince even an amateur that the animal could not have climbed trees; the shape and proportions of the limbs and the orientations of the joints are all wrong. In its adaptations for running, the limbs are jointed to swing only in the fore and aft plane. They cannot be rotated or flexed sideways. Indeed, it is well known that you can never be kicked by a horse if you stand well off to its side; a horse simply cannot swing its leg out sideways. Nature always has a few ringers, of course—like ducks that nest in trees, kangaroos that climb trees, or fishes that walk on land. In fact, fishes alone are so diverse that we can point to the climbing perch, the flying fish, and the walking catfish. One has to be very careful in applying arguments from physical principles and the description of morphology; corroboration is always needed from other lines of evidence.

We can reconstruct a lot of the behavior and even physiology of an organism from incomplete evidence, such as the skeleton or a fossil, without ever seeing the whole living organism or being absolutely sure we are right. Putting it more scientifically, we can make a number of hypotheses and hope to test them more fully as the direct evidence slowly comes in. Strictly speaking, you can never "prove" hypotheses; what scientists actually do is to try to amass more and more data that will turn out either to corroborate or refute their hypotheses. This makes science a lot of fun. The scientist builds up his experience and expertise and then puts himself on the line with an hypothesis. There is no great stigma about being wrong if the argumentation and reasoning were appropriately rigorous. To see an hypothesis confirmed is a thrill. As the following pages will show, there have been many hypotheses about the biology of coelacanths. We are slowly discovering which ones were wrong.

CHAPTER 6

Where
Do They
Live?

My complete disbelief in the
"inaccessible depths" idea did not, of
course, solve the problem.

—*J. L. B. Smith*

In this last decade of the twen-
tieth century we should be able to pinpoint the home range of
a group of six-foot-long, 150-pound fishes. That we may not
yet completely know where they live is a sign that the world is

still a large place. It increases our respect for the fact that vast regions of its oceans are still unexplored fully. On the other hand, perhaps we already know the answer; perhaps coelacanths really do live only off the Comores.

Certainly it is the consensus among ichthyologists that *Latimeria* is to be found only off the Comores, and that the South African specimen was a stray. But we know that most marine fishes, especially large ones, usually have very wide geographic distributions. Small shore fishes may be limited to one region, particularly if they are island fishes. But all the other large fishes, including the sharks, deep-water mackerellike fishes (Scombridae), or basses that live in the same Comoran waters as *Latimeria,* are found all over the Indian Ocean and beyond. If the living coelacanth really is restricted to the Comores, there must be some very special reason.

The question, Where do they live? is, of course, bound up with the bigger question, why do they live there and not somewhere else? *Latimeria* is not distributed worldwide. Does its apparently limited distribution relate to the fact of its survival, instead of extinction, at the end of the Cretaceous and the complex geological history of the western Indian Ocean. What evidence do we have on these subjects, and how shall we interpret it?

We have already seen the evidence of the catch data. Since the French survey was completed in 1972, perhaps another 100 or 150 specimens have been caught, and there are no surprises: *Latimeria* still is caught only off two of the Comores (Grande Comore and Anjouan) and only at night. Of the Grande Comore catches, the great majority have been from the west coast. There are somewhat more catches on average between December and March than the rest of the year, and they are usually taken at depths of one hundred to three hundred meters, at one to three kilometers from shore. But there is always the case of the single fish, that very first fish, eighteen hundred miles away to the south of the Comores—in roughly

the same depths, at the same season, and also at night. It was a
big one, and a male. So where do they live?

The reader will probably already have spotted the principal
message of the catch data. Ideally the catch data would give us
a perfect digest of Comoran folklore information about the
distribution of *Latimeria,* and we would expect that information to be pretty accurate. However, the data are not really a
direct reflection of *Latimeria*'s biology; instead, they tell us
about the activities of the *fishermen.* All the principal Comoran
fishing villages from which anyone fishes the deeper island
slopes are on Grande Comore and Anjouan. There are no
villages on Mohéli and Mayotte from which fishermen regularly work deep longlines, and almost no such fishing is done
from the east coast of Grande Comore. When the fishermen go
out with deep lines, they usually fish at depths of seventy to
two hundred meters. And their fishing is mostly confined to
the months of December to March. When they fish at depth,
they fish at night.

Why do Comoran fishermen show this pattern of fishing?
Certainly not because they have been fishing deliberately for
coelacanths (at least not until very recently). They fish from
December to March because that is when the weather is best,
and indeed, for the rest of the year it may be too stormy to
venture out far on the ocean in a dugout canoe. They fish at
night and at those particular depths because they are seeking
another, quite different fish—*ngessa,* the oilfish *(Ruvettus pretiosus).* This fish, as its name suggests, has an oily flesh that
yields a medical oil that has a strong purgative effect *(Ruvettus*
is sometimes called the castor-oil fish) and is also used as a
salve against mosquitoes and other insects. *Ruvettus* is also a
large fish, and even though not abundant, it is very valuable,
well worth the development of a specialized deep-line fishery.
It was the *Ruvettus* fishermen who caught *Latimeria*—but by
accident.

So the raw catch data really tell us more about the behavior
of fishermen who are trying to catch a different species—*Ruvettus pretiosus*—than they tell us anything directly about the ecol-

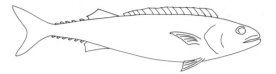

FIGURE 28   *Ruvettus pretiosus.*

ogy of *Latimeria.* But at the very least the data tell us about the overlap between the ecology of *Ruvettus* and *Latimeria. Ruvettus pretiosus* is a fish that is distributed very widely in the Indian and Pacific oceans. It is caught at depths between one hundred and three hundred meters. There are three places where a specialized *Ruvettus* fishery using deep longlines has been established: the Comores and the Cook and Society islands in the South Pacific. The fact that *Latimeria* has never been caught at the other two places where *Ruvettus* is regularly taken tells us (negative evidence though it is) that the distribution of *Latimeria* may be more restricted.

Where the coelacanth lives is really a double question. We need to know, Where horizontally?—the geographic distribution—and Where vertically?—shallow or deep in the water column. Let us first examine the question of depth range that so vexed Smith. The capture data show that all but two of the known coelacanth specimens have been caught at three hundred meters or less. But we may have to qualify that statement. First, no one has fished deeper, so they may be common farther down. Second, the estimates of depth come from the local fishermen, and we can be fairly sure that some of the reported depths have quite innocently been exaggerated, so they may actually be living in more shallow waters. Fishermen usually report the length of line payed out, which is always more than the actual fishing depth. Once any length of line at all has been set out, currents and the drift of the boat will bend the line into a long catenary curve. When an expedition from the California Academy of Sciences checked depths reported by local fishermen against depths measured by divers, they found major overestimations. A French oceanographic team found the same thing.*

The fishermen seem rarely to work at depths below about three hundred meters. The effort involved is too high, and the yield in *Ruvettus* may not require it. The investment in a six-hundred-meter line, the risk should the line be lost, and the simple difficulty of working such a long line by hand are too great for it to be general practice to fish at such depths. All this means that we simply have no idea how deep the range of *Latimeria* extends.

In addition to the direct evidence, we can call upon the

*However, we should not dismiss all local information as suspect. When the French authorities sent me information about the capture of the Yale frozen specimen, they were aware of the problem, noting that the actual length of line used by the fishermen was 220 meters and that the depth of capture should be estimated as between 150 to 200 meters. The latter figure is what Millot, Anthony, and Robineau recorded in their survey of catch data. Fricke found his specimens between 117 and 198 meters and, of course, measured the depth directly from his submersible.

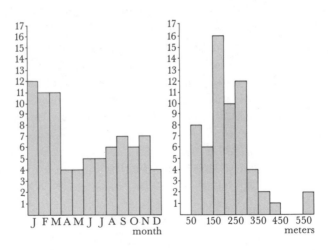

FIGURE 29   Distribution of coelacanths caught between 1938 and 1976 by depth of capture and by month of year. Data from Millot, Anthony, and Robineau and from McCosker.

biologists to play detective; we can infer information about the depths at which *Latimeria* lives from biological features of the fish itself. Right from the beginning Smith was sure, simply from the color of the fish, that it was not a truly deep-sea form—that is, it did not come from a thousand or more meters deep. All truly deep-sea fishes tend to be black in color (and, interestingly, the crustaceans tend to be red) but never blue. Bluish fishes always live within the photic zone (where light still penetrates). There is much more evidence along this sort of line. Deep-sea fishes are usually not as heavily armored as *Latimeria.* They tend to have enormous eyes and much larger mouths. The bones are light in construction. *Latimeria* has moderate-sized eyes, thick scales, and well-ossified bones. The scales and bones show growth markings indicating life in a seasonal environment. Smith concluded from all this that *Latimeria* was a fish of the deeper regions of coastal reefs, but not from below about two hundred meters.

Although their swimming seems effortless to us, fishes do not naturally float. That is to say, the tissues of the body of a fish—muscles, bone, blood, etc.—are slightly but significantly heavier than seawater (have a greater relative density). So all fishes must adopt strategies for controlling buoyancy. If they are not to sink and use energy to keep swimming, they need some special device to reduce their overall density. Many modern bony fishes use modified lungs as a gas-filled buoyancy organ or swim bladder. By controlling the volume of gas in the bladder, they can control the relative density of the body as a whole.*

In Chapter 1 we noted Marjorie Courtenay-Latimer's observation that coelacanth tissues are extremely oily. She and the taxidermist also discovered that coelacanths have a single

---

*Some fishes force air directly in or out of the swim bladder. Others can make themselves heavier or lighter than water by physiological action—secreting gas into the swim bladder or reabsorbing gas from it. These are useful adaptions because they can thereby achieve the correct relative density to be neutrally buoyant at different depths without access to the surface.

"lung." The French discovered that this modified lung is not an air bladder but is filled with oils and fats. Analysis by two research groups has shown that the fats and oils in coelacanth tissues are chemically complex, full of waxes and esters (organic chemicals in the fat family that have very low relative densities).[98] Because of all this oil, the general tissues and particularly the lung were slightly lighter than normal tissue like muscle and much lighter than bone.

As a result of having all this fatty material, coelacanths are just less than neutrally buoyant relative to "average" seawater. The coelacanth therefore has a passive buoyancy organ, which is rather like a diver's choosing the right weights on his or her belt for a particular depth. Within a certain depth range *Latimeria* can swim freely without having to worry about falling or climbing because of density problems. The question then is, Can we calculate the range of depths for which the special fattiness of coelacanth tissues is adapted? The answer from the organic chemists was: Coelacanths are adapted for maintaining station passively in the range between two hundred to three hundred meters maximum. Not surprisingly, considering its name, the oilfish *Ruvettus* uses exactly the same adaptation, and so do many sharks.

Another set of evidence comes from the eye. All the coelacanths that have been caught off the Comores have had something in common, a problem with the eye. The lens is cloudy when they are brought to the surface. In Fricke's movies the animals in their natural habitat have a good clear eye lens. It was realized very early on that the cloudy lens was caused by pressure change. As the fish is dragged up to the surface by the fishermen, rapid change in pressure alters the crystalline structure of the eye lens. These particular fishes, therefore, did not normally live in shallow surface waters.

Dr. Adam Locket (formerly of the Institute of Ophthalmology in London) is a keen student of coelacanths and participated in the 1972 expedition. He and his colleague Dr. H. J. A. Dartnall studied the eye of *Latimeria* and compared it with the eyes of other fishes.[99] First they studied the retina (the layer of photoreceptive cells onto which the incoming light signal is focused) to see what intensity of light it was adapted to receive. The French, observing a live *Latimeria* in 1954, noticed that it avoided the light and was seriously distressed in strong daylight. Studies on the fresh eyes of the 1972 specimens showed that the retina was adapted to receive weak light intensities in a range corresponding in seawater to a depth of up to several hundred meters. Comparative studies on other fishes put the coelacanth in company with the shark *Centroscymnus,* which lives at the depths where *Latimeria* is caught, and, to nobody's surprise, *Ruvettus pretiosus.*

Underwater, red and yellow colors are soon absorbed by seawater; the predominant colors of light are blue and green. Locket and Dartnall's results showed that *Latimeria* is essentially color-blind. The retina of the eye almost completely (but not quite) lacks the sort of color receptor cells called cones, and the cells that are present (rods) are adapted to receive light intensities and frequencies typical of fishes living at a hundred to two hundred meters, rather than in more shallow waters or the deep abyss. So the evidence of *Latimeria* eyes confirms the evidence of the body color and the fat content.

There are two theories about the vertical distribution of *Latimeria.* One is that it actually lives where it has been caught; the other is that the fish migrates vertically, like some of the creatures that it evidently feeds on. The main reason for wanting to believe the second theory is that the fish is always caught at night, and many people, starting with E. I. White in 1939, wanted to explain the discovery of *Latimeria* as the chance capture of a rare deep-sea fish. In some way it was easier to believe that the fish normally lives at great depths out of the normal range of science than to believe that there are still largely unexplored parts of the world's surface waters. There is, however, no good scientific reason to accept the migration theory in the sense of a really extended migration from the deep ocean and back on a daily basis. Dr. Fricke followed individual coelacanths with his submersible and found that they drifted passively with the current, making only a very modest migration between roughly two hundred and one hundred meters.

A particularly nasty problem for fishes is water temperature. Deeper ocean water is usually much colder than the surface water. Early measurements off the Comores showed that a fish living at 200 meters would be in water of about 15° to 18° C. At 500 meters the temperature is around 10° to 11° C, while the surface water could be as high as 27° C. In general the waters around the west coast of Grande Comore were colder than the east. An important question, therefore, is, What is relationship of the distribution of *Latimeria* to water temperature? Fricke found that the fish he tracked stayed below the 18° C level.[100] This is consistent with the general catch data: Fricke's fish stayed below about 170 meters, except where upwelling currents at night brought cooler water nearer the surface. In that case the fish drifted with the current to depths of around 100 meters. Fricke speculated that the coelacanths rest during the day in the cooler, deeper waters that suit their metabolism and then feed in the warmer, more productive waters nearer the surface at night.[101]

With all our information put together, biology and the catch data agree: *Latimeria chalumnae* seems to be a fish adapted for life between one hundred and five hundred meters.

All this brings us to the second problem: geography. Do coelacanths live only around the Comores? Again, direct evidence is very limited: 150 to 200 specimens caught from the Comores and just one from over a thousand miles away. And we have no idea about elsewhere. We can boil everything down to four questions. First, are the Comores the sole territory of *Latimeria*? If so, the Chalumna River specimen was a chance stray. Second, if this is true, we have to ask, What special qualities of *Latimeria* and the Comores explain this restricted distribution? Third, if *Latimeria* is not restricted to (centered on) the Comores, where else do they live? All this is bound up with the fourth question: Because the Comores are geologically very young, where did *Latimeria* live before the Comores existed?

The principal line of argument suggesting that the coelacanth *Latimeria chalumnae* might indeed be restricted to the Comores is theoretical. It comes from the study of paleobiogeography, the study of ancient distribution patterns of animals and plants and their changes over time. Darwin pointed out a consistency about relict species of the sort that we call living fossils: They generally have very limited geographical distributions.* Relict species are generally thought to be at a competitive disadvantage with respect to more advanced forms and manage to survive only by retreating to some isolated environment where some special circumstance allows them to hang on. Therefore, in principle one would not expect a "relict" species like *Latimeria chalumnae* to have a broad geographical distribution. (In which case, for an oceanic fish, what do you call broad—hundreds of miles, or thousands?)

This weak argument would be knocked on the head by the discovery of a single specimen from somewhere else in the world, and we already have coelacanth number one from South Africa. However, if it is correct, it would mean that the original stocks of *Latimeria* have become restricted to a population biologically incapable of living anywhere except on or around deep submarine slopes and by some accident geographically confined to the region of the Comores.

For a fish specialized for living on the deeper slopes of coastal margins, truly open water—typical deep ocean water—would actually be a barrier to dispersal. The fish would rarely, if ever leave, the safe haven of the Comores. Even so, individu-

---

*Among classic vertebrate examples zoologists always cite in this regard the three genera of surviving lungfishes, which are restricted to particular river basins in Australia, South America, and Africa.

als would still occasionally be swept away from the safety of the island margins. Which way would the surface currents of the region take them? The great westward-setting South Equatorial Current divides to flow around Madagascar. The westward arm of this current—the Mozambique Current—flows southwest down the Mozambique Channel. Thus we would expect to find that either (1) the Comores are the most southwesterly extension of the coelacanth's range and perhaps even the Comores' fishes are outliers from other populations out in the general Indian Ocean to the northeast, or (2) coelacanths have their main center in the Comores but could extend southwest from the Comores all down the two sides of the Mozambique Channel.

The 1969 expedition tested the first possibility. The banks immediately to the east of the Comores should be an ideal place for *Latimeria*. The islands of the Aldabra-Astove group to the north are perhaps less likely. No coelacanth was found. With respect to the second alternative, one specimen *has* been found far down the Mozambique Channel. One specimen out of two hundred plus is not enough to make a case.

All in all, one can make a strong argument that *Latimeria* exists principally in the Comores, probably because of special physical and biological conditions that either directly favor *Latimeria* or reduce the level of potential competition and predation. The South African specimen, in this case, is a stray, but such strays are likely to be found, if only rarely, southward along the Mozambique Channel.

Next we have to deal with the negative evidence. If any other populations of *Latimeria chalumnae* exist, how big would the population have to be, at what depths might it have to live, and how wide would its geographic range be, in order to have a chance of detection along, say, the coast of mainland Africa or on the other islands of the western Indian Ocean, including Madagascar? In other words, how reliable is our *negative evidence* that they haven't been caught anywhere else?

Other things being equal (of course, they never are; for

example, one cannot know whether coelacanths have a special taste for some trace metal found only in the Comores' volcanoes), one would say that the coelacanths *ought to be able to live anywhere* around similar islands in the western Indian Ocean. *Ruvettus pretiosus,* for example, occurs throughout the whole Indian Ocean and in the Atlantic and Pacific. None of the other large fishes caught off the Comores seems to be confined to the Comores. And no other island group in the western Indian Ocean has unique (endemic) large species of its own although there are some smaller endemic fishes, confined to shallower reefs and lagoons, that have very limited distributions.

Smith thought that the negative evidence became very strong as the years went by. He argued that trawlers fishing the South African coast would by now have caught another coelacanth. Sufficient fishing effort had been put in, at least on the mainland coast from Cape Hope to Kenya, to conclude that coelacanths do not occur there. He thought the same was true of Madagascar. For the other islands of the western Indian Ocean, one can state only that the verdict is not yet in. Each year without catches from elsewhere (rather weak evidence to be sure if no one is *actively* looking for *Latimeria* anywhere else) the idea gains that the living coelacanth occurs only on the Comores gains more and more strength.

Despite this, several authors (Fricke and Michael Bruton, for example) have continued firmly to believe that the East London specimen is not a stray and that an African coastal population must exist.[108] Until we know a great deal more about the sorts of fishes that live below 150 meters over the whole Indian Ocean, any conclusions must be tentative. We must always remember T. H. Huxley's famous maxim that it takes only one ugly little fact to destroy a beautiful hypothesis.

What is special about the Comores? They rise up almost vertically from the ocean floor as the cones of volcanoes, and Mount Karthala is still active. The islands rise up into the southerly-flowing Mozambique Current and are separated from both Madagascar and the African mainland by water reaching three thousand meters deep. It is a characteristic of oceanic volcanic islands to have steep submarine profiles. Immediately offshore, water depths increase very sharply, and the bottom may slope at angles from twenty to forty degrees.[103] To put it another way, the richly productive surface water that one expects to find on any coast is here restricted to a very narrow band, compared with the broad shallow shelves surrounding most continental masses. But on most comparable islands in the western Indian Ocean, there is a broad fringing coral reef system within which fishes teem in the productive sunlit shallow water. One special feature of Grande Comore and Anjouan is that they lack a complete fringing reef.[104]

Because they are large mountainous islands, the Comores have a significant rainfall that quickly soaks into the porous volcanic rock. This freshwater finds its way through underground aquifers and is released into the ocean along the submarine slopes. It has been a favorite conjecture of some zoologists that this influx of freshwater provides a special localized environment in which the coelacanths live. This sort of undersea freshwater outflow is not unique to the Comores, of course, but it might be especially well developed there. However, Fricke found no freshwater outflows below eighty meters, and the water where *Latimeria* is caught is quite normal salt water.[105] All the physiological evidence argues strongly against this romantic hypothesis.

With recent studies that have given us accurate ages for the Comores, we can add a big new factor: Grande Comore is also the most recent of the islands in the chain. Anjouan is dated at more than 1.2 million years while Grande Comore is only 130,-000 years old. Several authors have noted that many catches from Grande Comore were from near recent larva flows on the submarine slopes. Perhaps young rock provides a unique set of underwater conditions, in either its chemistry or simply because the lava has not had time to develop a flora and fauna to sustain a normal biological productivity. The whole environment is different from that of other oceanic islands, starting with the absence of a fringing coral reef system.

The lack of a fringing reef strongly affects the fishing habits of the inhabitants and may in fact be the reason that the fish was discovered at all on the Comores. If there were a decent inshore reef environment, the fishermen would probably never have ventured beyond it; the medicinal value of *Ruvettus* set aside, it would be safer not to bother with that fish. Somewhat corroborating all this is the result from the 1972 expedition. In his experiments with vertical longlines, using the same techniques developed for use (with great success, except no *Latimeria*) off the islands to the north in 1969, Forster discovered that in the waters of the deeper slopes of Grande Comore fishes of any kind are very scarce.[106] In terms both of species diversity and simple numbers of individuals, fishes are perhaps seven to ten times less frequent than expected. The same species are found as from the other islands, but most of the species that one would expect to find are missing. Forster even failed to catch the large sharks like *Hexanchus* that abound in the waters down to four hundred meters off the islands to the north where we fished in 1969. Observing off Grande Comore in his submersible, Fricke found the same thing.[107] The fauna of fishes was poor and especially poor on the east coast.

What does all this mean? Perhaps only in an impoverished environment can a relatively inactive, sedentary fish make a living. This is a weak argument for explaining the special occurrence of *Latimeria,* but the differences in fish abundance and diversity between the Comores and the islands immediately to the north are a very real phenomenon and must surely be related to their very young geological age, the recency of the lava flows, and the absence of coral reefs.

Where did the ancestors of modern *Latimeria* live before the Comores existed? This is even harder to answer. There are no fossil records of coelacanths since the Cretaceous, when the closest relatives of *Latimeria* lived in Europe. The ancestors of *Latimeria* were probably fishes of shallow epicontinental seas, particularly of waters of less than five hundred meters, rather than of the open and/or deep ocean. Since the Indian Ocean did not exist before about 125 million years ago, the ancestor of *Latimeria* must have lived in the great sea called Tethys, a sea that in Cretaceous time was bordered by Eurasia to the north and Africa to the west. As the continents moved and rotated, this Tethyan sea gradually closed up, but the Indian Ocean opened.

The last 125 million years have seen a worldwide explosion of fish diversity and evolution. We can make the argument that the forebears of *Latimeria* might very quickly have become at a disadvantage relative to these new faunas and were forced into environments (including, but not exclusively, those around relatively recent oceanic volcanic formations) where competition from other large fishes was reduced and predators were few.

As the continents separated, the coelacanth stock could have migrated southward from Tethys into the opening Indian Ocean and along the coasts of Africa, Madagascar, and India. But safe depauperate environments were most likely to have been found on the new island chains forming in the Indian Ocean itself, from the west of India to the Comores. We have seen that these were formed in a series from the Seychelles, the Amirantes, and the Farquhar islands to northern Madagascar and then the Comores. The Mascarene Ridge has probably always been deeper than one thousand meters, but it forms a link to the more recent islands of Mauritius and Réunion. To the north and east coelacanths could have followed a separate and parallel route across the large chain of islands—the Laccadives, the Maldives, the Chagos Archipelago—formed at the mid-ocean ridge system.

This gives us a rather melancholy picture of *Latimeria*'s forebears being pushed from island to island across the Indian Ocean as environments successively matured and fish faunas diversified. There are few places where they could exist now.

Having come along the Seychelles to Comores chain, they might perhaps have reached Réunion via the Mascarene Ridge; the fauna of Réunion is not well known. They might also have reached the Maldives and Chagos Archipelago, but these do not have depauperate fish faunas or extensive recent vulcanism and therefore probably could not provide the right habitats.

Putting all this together, I conclude that the living coelacanth may well be a true relict species, existing in any numbers at all only on the Comores, but that if it does exist elsewhere, it would be in the form of stragglers from Comoran populations carried to the deeper slopes of the Mozambique coast and farther south, as Smith predicted, but where faunas are depauperate, predators few, and perhaps volcanoes have recently been active. If any other population of coelacanths evolved in parallel to the Comoran stock, it would most likely again be found on geologically recent island with a depauperate fauna to which the stock could readily have migrated from nearby—that is to say, at the end of an island chain. After Bassas da India and Europa in the Mozambique Channel, the next place I would look, therefore, for more living coelacanths would be on Réunion.

The Mediterranean?

Strangely enough, we cannot stop here with the scientific evidence alone. A fascinating footnote to all this has cropped up in the world of ethnography. A priest from a small church near Bilbao, Spain, is reported to have sold to an Argentinian chemist in 1964 a small votive figure of a fish, made in silver and possibly nineteenth century in origin.[108] To many who have seen the photographs of this object it looks for all the world like a model of a coelacanth. Later a second silver model of the same sort from Toledo was found in a Paris antique shop.[109]

Votive figures of fishes are not uncommon, especially in

fishing communities. One would expect the craftsmen and those who commissioned the figures as a prayer for good fishing to know their fishes quite well. It would be one thing to make a stylized fish combining features of known food fishes (sardine, tuna, herring, or cod), but such a concatenation would scarcely produce the lobed fins and trifid tail of a coelacanth. So what is the origin of these figures?

There are only three possibilities: These figures are meant to be representations of coelacanths, the similarity is pure chance, or the figures are fakes.

If these figures truly represent coelacanths, the artisans must have seen specimens or descriptions of specimens of a coelacanth, and there must have been some special reason to make such a votive figure. Of course, the artisans could have seen a fossil coelacanth. The fossils readily show a trifid tail and two dorsal fins. However, fossil coelacanths very rarely show the lobed structure of the paired fins. The reason for this is that the fleshy contents of the fin lobe are subject to decay, and in that process the bones of the internal fin skeleton are destroyed, leaving a gap. Only an expert would have known that something had been present in that gap. It is unlikely, therefore, that the figures could be modeled after fossils.

Fricke believes that the artisans had seen a real coelacanth specimen and that populations of *Latimeria* or some related

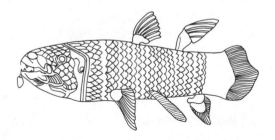

FIGURE 30   Sketch of the silver icon from Toledo, Spain.

species must recently have lived in the Mediterranean or Red Sea and might be there still.[110] This is a romantic notion for which there is not a shred of evidence. The fishes of the Mediterranean and Red Sea are very well known by now, and no coelacanth is among them. (Bilbao is on the Atlantic coast of Spain anyway, and no one thinks an Atlantic population is possible.) However, it is always possible that someone trading along the African coast might have acquired a coelacanth, salted and dried perhaps, and found it interesting enough to bring back to Europe. In that case the votive figure might have been made in the hopes of furthering good fortune in other African trading ventures, rather than as a talisman for good fishing locally. Professor Anthony believed that the artisan(s) might have visited the Comores.

The simplest explanation is that an artisan merely made up a fanciful fish and by chance produced something looking a bit like a coelacanth, with the rest being in the eye of the beholder. Then there is always the possibility that we are dealing with a fake. More subtly, it is possible we are allowing ourselves to see more than is really there. For example, in the case of the first figure, photographed and published in *Sea Frontiers* in 1966, while most of the fish is quite realistic-looking, the pectoral fins look nothing like the fins of any fish except possibly a mudskipper *(Periophthalmus)* and seem to have been added rather crudely. The second dorsal fin is not a lobed structure, but a rayed fin exactly like the first dorsal. In the Toledo specimen the paired fins are less dramatically emphasized and the tail is rather undefined. Both have accentuated scales, but that is typical of depictions of fishes made by untrained persons.

It is remarkable that scientists are willing to throw their normal dispassionate caution in cases like this, especially when dealing with evidence from another sphere such as art. We could call this the Loch Ness Monster phenomenon, something that would be wonderfully glamorous and exciting if it were true. However, what is fine for journalists is scarcely science. A cold, hard look at the facts should tell us the following:

These icons may not represent coelacanths at all, and even if they did, they do not constitute evidence that *Latimeria* or any other coelacanth species is currently living in the Atlantic, Mediterranean, or Red Sea—much as we would like that to be the case.

How They Live: Swimming and Feeding

[A] nocturnal piscivorous
drift-hunter.

—H. Fricke

Some years ago I lectured at the Mount Desert Island Biological Station at Salisbury Cove in Maine, a delightful place where medical researchers gather each summer to study comparative physiology, mostly in connection with kidney function. A lot of work is done there on the spiny dogfish *(Squalus acanthias),* a small, common, and rather homely shark. I teased those in the audience by telling them they did not work with real sharks and showed them a picture of a magnificent streamlined blue shark *(Prionace glauca),* swooping toward the photographer (not me), leaning hard on its pectoral fins as it made a broad turn, its tail driving powerfully to one side. With its mouth slightly open, showing the teeth, it was a superb machine for swimming fast and killing. "This is a *shark,"* I claimed. But it was unfair, because the smaller sharks living near the bottom of the colder North Atlantic waters and many others are just as graceful in the water. For my money, to understand a fish, you have to watch bigger species like these, the shark, salmon, bass, bluefish, marlin, or tuna. The sorts of fishes that live in home aquariums tend to hover in one place most of the time, darting here and there just to let you know they are keeping busy. They are beautiful, too, and, if you study them closely, really fascinating. But for the gestalt of being a fish, a larger ocean fish is the place to start.

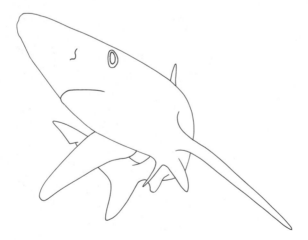

FIGURE 31   Blue shark in tight turn.

Over many years I have spent hours simply watching fishes and have found that the best place for that is the large public aquarium. Here, while the fishes slowly circle, you can see the

exquisite control they have over their movements in three dimensions. The tail thrusts the animal steadily forward, using
what seems a simple action. But all the fins, even if only right at
their very tips, and the edges of the tail itself, are constantly
adjusting delicately to the flow of water over the body in order
to produce this totally controlled action. The bird books always describe the eastern kingbird (*Tyrannus tynannus*) as flying
like a butterfly on the tips of its wings. Fishes, even more so,
cruise with the power of their trunk muscles but on the tips of
their fins. Then, when a burst of speed is needed, the fish's
body visibly stiffens, the fins are drawn in, the side-to-side
bending of the tail is reduced in amplitude, and almost with a
series of shivers the fish accelerates smoothly away.

Perhaps more than any other kind of animal, fishes demon-
strate in form and appearance the life-style to which they are
adapted. And the range of adaptations is enormous. Many bigger fishes, like sharks and tuna, swim with their mouths open
so that their swimming helps power water into the mouth and
over the gills. Smaller fishes, hovering in weeds or rocks, actually find that the action of pushing water over the gills drives
them slowly forward by jet propulsion, and so they must fan
their paired fins, like little helicopters in the water, in order to
counter this and remain still. (Sharks cannot swim backward;
score one for the aquarium dwellers!) In a curious set of evolutionary shifts, sharklike fishes gave rise to flattened rays and
skates that live on the bottom and swim by rippling the edges
of the large body disks (formed of expanded pectoral fins).
These in turn gave rise to free-swimming forms like the man-
tas and cow-nosed rays that swim through the water using expanded pectoral fins like nothing more or less than great
flapping wings. There is something almost sinister about the
silent flight of a ray through the water, but it is also supremely
beautiful, like the flight of an albatross or a condor. Among the
ray-finned fishes, exactly in parallel, there evolved other
groups of bottom-living fishes, the flounders and their relatives, but these became flattened by lying over on their sides.
This would have put one eye in the mud, so the whole head

was grotesquely twisted around at the same time, and this met-
amorphosis is repeated in the early embryonic life of each individual. The range of such morphological adaptations for different patterns of swimming and different modes of life among
the fishes is almost infinite in variety. In among all this diversity of fishes, graceful and grotesque, where does *Latimeria* fit?
What kind of life does it lead on the deep reef faces of the
Comores?

SWIMMING

The simplest aspect of the biology of *Latimeria* to decipher may
be this matter of locomotion. Moving in water is a mechanically highly constrained function. You can tell a great deal
about how a fish can swim from its shape. Torpedolike fishes
cruise in hunt of their prey; fishes that live on the bottom like
flounders or skates tend to be strongly flattened; fishes that
can swim at high speeds tend to have sickle-shaped tails (some
sharks and tuna, for example); fishes that hang around vegetation and other cover have shorter, deeper bodies and large,
feathery fins (coral reef fishes, goldfish, Siamese fighting fish,
for example).

The biologist's job here is to take the grace and power of
swimming and explain it in terms of ordinary mechanics. Any
kind of swimming (and, indeed, all movement) depends on
Newton's third law of motion: To every action there is an equal
and opposite reaction. To move forward against the water, the
swimmer pushes or pulls at it, and the water returns the compliment, pushing the body forward. In walking or running, you
push hard backward and the ground pushes equally hard on
your leg forward. Since you are movable and the ground is not,
you move forward. On ice or soft sand, where the transmission
of force between you and the ground may be broken through
loss of frictional contact, the tables may be turned. Water can
be caused to flow, but it is incompressible (at least with the

forces fishes can generate), so water readily "pushes back," making swimming possible.

Vertebrate swimming began in the Cambrian with small elongate animals that could flex their bodies from side to side, their flanks making an inclined plane that pushed sideways and backward at the water, the water pushing sideways and forward in return. All vertebrates use the same mechanism. The muscles of the trunk in fishes are organized in a series of discrete muscle blocks (segments) arranged in pairs along each side of the body. A wave of muscle contraction fires down each side, so that the segments contract in sequence, one after the other. The waves of contractions are exactly out of phase on the two sides of the body, so that the body is bent into a set of curves. As the waves of contraction repeat, in exactly opposite phases on either side, the body is constantly flexed from side to side, to push against the water.

The system then depends on the muscles' being in discrete units and firing in sequence, out of phase from side to side. If they all fired in unison, the body would simply lock up. But something else is needed. All the muscle fibers in these blocks of swimming muscles run parallel to the long axis of the body so that when they contract, they tend to shorten the body at that region, forcing the body into a bend. This means that when the muscles contract, they all are shortening in the same direction. What then prevents the whole fish from squeezing up like a concertina? The answer is that this was originally the all-important function of the notochord. It acts as a longitudinal stiffener, a brace against which the muscles can bend the body without collapsing it. The notochord also stores energy at each bending and then releases it on the return stroke, for greater efficiency. (In the very earliest chordates the notochord was actually a kind of elastic muscle.)*

---

*In a fish like an eel, this S-shaped bending of the body is very pronounced and the tail fin is not very large. In advanced high-speed fishes like the tuna and marlin, the tail is sickle-shaped and set off from the body by a narrow stem

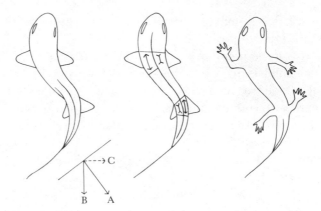

FIGURE 32   Swimming and walking originally depended on body flexure in an S-shaped curve. Note the direction of the muscle fibers in the block of muscles (segments). As the tail pushes sideways against the water (A), there is a backward component of the force (B) to which the water responds by pushing the fish forward and a lateral thrust (C) that is essentially wasted.

Fins were added to the vertebrate body plan very early in evolution as control surfaces. The median fins and particularly the large dorsal fin are used to provide lateral stability in the water. They counteract the tendency of the tail to send the fish off course and to prevent the fish from leaning over sideways. The paired fins probably first arose in fishes as devices to aid in braking and turning. The fish sticks out the pectoral fin to increase drag on that side and to slow that side down. The fish then slews around, using the fin almost as a pivot. Many fishes have large paired fins that are used in rowing actions for slow swimming. The paired fins of a shark, however, are solid and stiff, resembling the diving planes of a submarine or the tail assembly of a jet fighter.

or peduncle. In these fishes the body itself does not move much from side to side. All the power is generated by the tail portion, acting like a powerful propeller.

The coelacanth, with its deep, chunky trunk and broad, blunt tail, has the general body type of a fish that *can* swim fast but only for short periods. It is the build of a fish that waits for its prey and then catches it in a short, sharp burst. It is a well-known principle of swimming that the force generated by the tail is proportional to its surface area and to the square of its velocity (from side to side). Unfortunately the drag that is produced by the fish's moving through the water is also proportional to the area and the square of velocity. Therefore, all big, fast-moving fishes must be very well streamlined. *Latimeria* is big and nicely shaped, but it does not have the refined lines of an oceanic shark or marlin. In general one should therefore expect (indeed, it was long ago predicted from mechanical principles) that the coelacanth, like a pike or a grouper, would normally swim around slowly, using its paired and median fins, particularly the two backwardly pointing median fins (second dorsal and anal) as sculls, slowly pushing itself through the water. The first dorsal fin is obviously for maintaining vertical stability. In fast swimming it would be folded back out of the water flow in order to reduce drag, but in hovering around and making turns, the first dorsal fin would be fully flared.

All early observations of *Latimeria* emphasized the extraordinary mobility of the paired fins. In the Yale 1966 specimen we measured that they can be rotated top edge forwards (pronated) through more than 200 degrees and backward (supinated) through 100 degrees.[111] One early French specimen was actually preserved with the left pectoral fin rotated through 180 degrees, and it sits against the side of the fish looking so perfectly normal that several scientists have been deceived into thinking it is the regular orientation.

The specimens that have been caught and kept alive briefly confirmed all the early conclusions about the swimming capacity in *Latimeria*. The French observed "curious rotating movements" of the paired fins of the 1954 specimen.[112] The fish taken in 1972 made lateral sculling movements with its second dorsal and anal fins and kept its first dorsal flared as it fought for balance.[113] This second fish kept its paired fins flared out and seemed to use them in relatively small movements, also for balance. It used the tips of the fins to touch the bottom, apparently for balance. However, Dr. Fricke, with his films of coelacanths *in situ* has dramatically extended what we know about coelacanth swimming.[114] First, as predicted, he did not find any of them engaged in steady fast or even moderate speed, cruising behavior. Instead, he saw them hovering near the bottom, keeping station in a current or very occasionally swimming in short bursts. When the fish hovers, it does use the second dorsal and anal fins in a side-to-side motion, rotating at the same time as if to maintain a fine tuning of the balance. And it does keep the first dorsal flared, but the principal swimming organs at slow speeds or for hovering in a current are the paired fins, which sweep to and fro in a rotating sculling action. When the coelacanth is in this slow-moving or hovering behavior, all the fins are fully extended, making the fish look very beautiful, even delicate, for all its size. Fricke's observations on swimming are slightly different from the observations made in 1954 and 1972 on line-caught fish released exhausted into shallow water or an enclosure. These latter fish did not use their pectoral fins much in swimming, but rather for balancing, and they used their second dorsal and anal fins for sculling slowly forward. This being the case, we have further reason to conclude that the 1966 fish photographed by Stevens, which apparently showed a similar behavior, had indeed been caught on a line and released.

Zoologists originally thought that the paired fins of coelacanths and the fossil lobe-fins functioned as true limbs, as props to lever the fish against the solid substrate of the bottom sand or against rocks.[115] The lobed fin base, characteristic of the whole lobe-fin group, consists of a central axis of skeletal elements and attached muscles and certainly looks quite strong enough, especially if we allow for the fact that the fish is weightless in water. So the view had built up that the coelacanth must spend part of its time on or near the bottom and must use its paired fins for an action akin to walking (but not, of course, having to bear any real weight). This view was

FIGURE 33   Sketches of swimming *Latimeria*, based on published photographs taken by Hans Fricke.

strengthened by the fact that the Australian lungfish *Neocerato-dus*, which has similar looking lobe-fins, appears to use them in this way.[117] Smith, in an uncharacteristic lapse into cuteness, called his book on *Latimeria Old Fourlegs*, further reinforcing this idea.

In fact, in considering the origins of tetrapod limbs, zoologists did not pay enough attention to the paired fins as fins; too

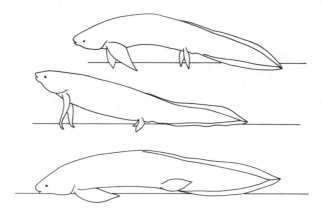

FIGURE 34   An Australian lungfish on the bottom, using its fins to balance and perhaps "push off," but not bearing any weight on its fins. After Dean.

much thought was given to interpreting *Latimeria* as a proto-tetrapod rather than simply as another kind of fish. When he filmed specimens *in situ* near the bottom, Fricke did not see the paired fins being used to "walk" in any active way—to prop the fish up off the bottom or to push against it. Here is another area where much still remains to be discovered.

Fricke's films did confirm the interesting fact that the paired fins are moved in a more or less regularly alternating pattern. That is to say, the two fins in each pair and the fins of the same side move in opposite directions. The pattern was not perfect, but to Fricke it seemed the same pattern of limb movements that one can see in the locomotion of any tetrapod (verte-brate).[117]

The alternating limb movement is clearly visible in a sala-mander, a lizard, dog, and a crawling baby and even in the way in which we swing our arms when we walk or run. It is obvi-ously deeply ingrained in tetrapod vertebrate locomotor me-chanics. However, it is not unique to tetrapods but is a conse-quence of the way fishes swim by bending their bodies. The fins are best placed at the node and antinode of the body's sine wave flexure. This pattern was set in place with the first verte-brate that swam by lateral undulation, and it is seen in some form or other in all vertebrates, including ourselves. (Only a very few tetrapods naturally use the paired limbs in more sym-metrical ways. A frog uses its hind legs together in swimming, of course, and horses and giraffes when "pacing" move the two limbs on a given side of the body in the same direction at each stride.)

Therefore, we have to conclude that the fact that coel-acanths seem to "swing their arms and legs" in the same alter-nating pattern as tetrapods is not a sign of direct relationship between the two but simply a reminder that both are verte-brates. To the extent that it is useful in the hovering swimming of *Latimeria* or the Devonian lobe-fins, it may be called a pread-aptation—a feature that functions perfectly well in one set of environmental conditions and then turns out to have an unsus-pected new functional significance in a second set of condi-tions. Lungs, which arose as a device for breathing when in the water, were preadaptations for breathing on land, for exam-ple.

The few modern ray-finned fishes that move around out of the water walk in a fashion quite unlike tetrapods; they lack the internal limb musculature. They have to use the paired limbs as simple passive props or crutches. The curious little mud-skipper *Periophthalmus* that lives in swampy regions of Asia, Africa, and Australia and the Asian walking catfish *(Clarias ba-trachius)* are good examples. Coelacanths and lungfishes have only a limited capacity to use the internal muscles and joints of the fins in a "tetrapod style" of pulling with the fore limbs and pushing with the hind limbs (Chapter 10), while they are much more mobile than tetrapods at the "shoulder" and "hip" joints.

Finally, a special feature of the tail in all coelacanths, living or fossil, is that it is completely symmetrical and there is a small separate terminal lobe. This median lobe can both flex from side to side (through almost 180 degrees) and rotate.

From observation of living coelacanths and from comparisons with other fishes, this terminal lobe seems to be a special trimming or balancing fin, adjusting for torque forces produced by the other fins that would tend to destabilize the fish as it swims or hovers and adding to the sculling function.

FEEDING AND BREATHING

Modern vertebrates arose from forms something (but not a lot) like modern lampreys and hagfishes—mentioned previously as living fossils from ancient radiations of primitive agnathan (without jaws) fishes. These Cambrian, Ordovician, and Silurian age fishes probably fed by scraping up vegetable matter, decaying animal remains, and small living animals such as worms from the bottom, into quite simple mouths. The modern hagfish feeds as a scavenger on the carcasses of dead animals on the seafloor. The adult lamprey is a parasite, holding on to live fishes via a suckerlike mouth and rasping away the tissues of its prey so as to feed on the blood.*

While the Agnatha lacked true jaws, they had well-developed gill systems. The gills are a set of respiratory pouches lined with blood capillaries for gas exchange and supported by a gill skeleton. In most fishes, gills are ventilated by muscles that pump water by means of compressing and expanding the gill and mouth chambers. The earliest forms may have had a simpler system using a skeleton of elastic cartilage so that the muscles came into play only in the constricting phase and there was a passive—elastic—recoil phase that followed.

The evolution of the complex jointed head of coelacanths from these primitive forms is surprisingly simple. The hypothetical ancestor of all the "jawed" vertebrates or Gnatho-

---

*The modern lamprey and hagfish have somewhat differently specialized rasping tongue apparatuses, and we do not know exactly when they arose. However, lacking biting jaws and teeth, none of these sorts of fish (living or fossil) could cut or grasp parts of large prey items.

stomata (jaw mouth) "invented" the first jaws by modifying the front end of the gill apparatus. The muscles and bones that worked the pumping mechanism for the first gill chamber became modified to squeeze onto and hold food items. This was the beginning of true jaws. The first jawbones (formerly gill bones) soon became covered externally with the beginnings of teeth. The old gill muscles became a powerful set of muscles for biting. This special history has meant, however, that the jaws and the gills of fishes were forever linked mechanically; they were, after all, part of a common system. And in all fishes the muscle system that opens the mouth by depressing the jaws is an integral part of the gill-tongue system.*

Tetrapod vertebrates, when they went out onto land and used the lungs instead of gills, finally broke the connection between feeding and breathing mechanisms—but not completely. We all have a relic of this history in the three little ear bones of the middle ear of mammals. There are actually three bones from the first two gill arches of our original fishy ancestors—two bones from the jaw arch and one from the first gill arch behind it. And the skeleton of the tongue and larynx is the final remnant of the skeleton of the gill apparatus.

After finding the prey, an animal has to get its jaws around it. This sounds simple, but the jaws do not operate like a pair of tongs or scissors. Very early in vertebrate evolution the whole jaw apparatus became firmly attached to the head (probably for added support as the power developed by the jaw muscles increased). While the lower jaw can be freely moved up and down (adducted and depressed), the upper jaws are stuck in place as a sort of anvil against which the mandibles

---

*A neat reminder of all this arcane anatomy is the human eustachian tube. One is not usually aware of this structure until one has a cold while traveling on an airplane. This tiny tube connects the middle cavity of the ear to the back of the throat. It works to keep the air pressure equal on either side of the eardrum. When you get a cold, the tube is blocked. The plane goes up, the pressure in the cabin goes down relative to the pressure in the middle ear, and the eardrum is stretched. This eustachian tube is nothing less than the remnant of one of the gill passages of our fishy ancestors.

strike. All vertebrates have the same problem. To take a piece of food symmetrically from above and below, a primitive fish would have both to drop the lower jaw (easy) and then to fold back the head in some way to elevate the whole upper jaw assembly. You can see this by what happens when you toss a bone to a dog or try to catch a peanut or candy in your own mouth. You have to rear your head back, and you can, because all mammals have a flexible neck. A fish doesn't have a neck because its gill apparatus is bunched up behind and underneath the head. It is even worse for sharks because their mouths are not terminal but slung beneath the head.*

To surmount this problem, fishes have evolved fascinating adaptations. Many advanced groups have shortened the upper jaws, moved them forward, and freed them from the rest of the head. The advanced ray-fins that do this include some truly remarkable deep-sea forms with jaws like mechanical hands. But such adaptations did not appear until the Late Mesozoic.

Now, finally, we can see what the intracranial joint of coelacanths is for. In Chapter 4 I mentioned that I had made models of how the intracranial joint might work in coelacanths and other lobe-finned fishes. According to my hypothesis of 1965, as the lower jaws drop, opening the mouth, the upper jaws and the whole anterior part of the head are raised by rotation at the intracranial joint. The result is that the mouth can be opened to approach a prey item from both below and above. The geometry of the coelacanth head means that the lower jaws are

---

*Imagine the difficulty if, for example, you were a fish living on the bottom with your throat pressed hard into the sand. How would you open your mouth? The lower jaw should be depressed, but it is already resting on the bottom. The upper jaw is blocked by the head unless the whole trunk can be raised up. For fishes swimming freely in the water column the problem can be avoided by coming up at prey somewhat from below. This is what sharks do. (Perhaps the little shark *Isistius brasiliensis* does it best of all. It feeds in a most remarkable fashion, approaching its prey [usually much bigger than itself] obliquely and grabbing on with its extremely sharp teeth. Then the whole body twists like a cookie cutter, neatly cutting out of the prey a cylindrical plug of meat.)

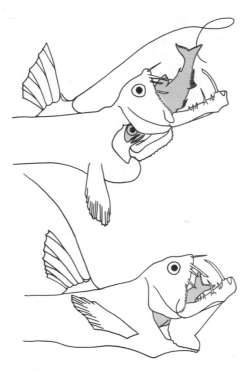

FIGURE 35 The anglerfish uses a fishing rod and lure derived from the first ray of the dorsal fin to lure prey toward the huge articulated mouth. After Tchernavin.

also swung forward at the same time. When the mouth is being closed, this is reversed, and the mandibles draw the food back into the mouth while it is enclosed from both above and below. It all forms a highly specialized feeding mechanism.

The only possible test of the hypothesis was to discover whether the joint actually moves in the only living fish that has the joint—namely, *Latimeria.* When I began the study, the conventional wisdom had it that the intracranial joint in all the living and fossil rhipidistians and coelacanths (including *Latimeria*) was quite immobile.

When we thawed Yale's frozen specimen in 1966, after everyone had left the laboratory, I had one last experiment of my own to make, a relatively simple one. But it was a tense moment; a prized hypothesis was on the line. There was no way I could avoid it after all the preparation. The fish was completely thawed and flexible. I grasped the tip of the snout (carefully, because the teeth are sharp) and lifted. . . . The joint hinged smoothly, the tip of the snout came up, and the lower jaw dropped and moved forward. There was no question about it. I felt a little like Galileo: "All the same, it moves." Then I repeated the actions, making motion pictures and using a series of markers attached on the head so that I could reconstruct the relative movements of the different parts precisely.[118]

In films of the living 1972 specimen, movements of the gill apparatus are well displayed because the fish was short of oxygen and the movements were exaggerated (they may for this reason also be slightly abnormal). It was possible to see a very faint movement of the tip of the snout during this breathing, but the fish never properly opened its mouth.[119] Fricke also did not succeed in filming any fish feeding, and this certainly

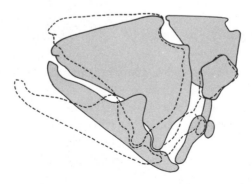

FIGURE 36 The simplest possible movement of the intracranial joint in *Latimeria* pushes the snout upward and the lower jaw forward. Much more complex movements are also possible.

remains a very high priority for me because as soon as we had established that the joint was indeed freely mobile, new questions were immediately posed. What is the movement actually for, and what muscles make it work?

These questions will remain unanswered until we can observe a live specimen feeding. Certainly the joint must help position the jaws around the prey. And we know that *Latimeria* uses the jaws to catch smaller fishes (six to twelve inches long). Coelacanths do not have massive teeth for crushing or a slashing jaw mechanism like sharks. We also know that they did not use blinding speed to catch their prey. The bite was more precise and controlled with the short, sharp teeth used for holding and positioning the prey. Perhaps the mechanism is used in part to create a suction mechanism that helps in prey capture. Perhaps the relative forward and backward movement of the lower jaws is critical. But it is a complicated feeding and respiratory movement; of that we can be sure. Moreover, we now have very good reason to believe that fossil lobe-finned fishes with similar joints also made specialized movements with their jaws and gill apparatus.

Which muscles operate the system? A neat pair of puzzles must be solved. We have to explain what pulls the anterior braincase region upward when there are no obvious muscles available on the back of the head. And we have to explain how the lower jaws can be driven forward when muscles can only contract and pull backward and upward. (Muscles can only contract; therefore, all muscles have to be arranged in "antagonistic" systems in which one pulls out and the other pulls in, as in the biceps and triceps muscles of the human upper arm.) A variety of ever more complicated explanations for that forward movement has been proposed. All of a sudden everyone had an opinion on this subject. First I proposed a mechanism, but it was far too simple.[120] Then in a major turnaround Robineau and Millot proposed a solution for movement at the joint, based on a small specimen (forty-two centimeters) that they obtained frozen from the Comores in 1973 (this was the first frozen specimen that the French laboratory had obtained).[121] But their account was wrong, and finally Dr. George Lauder seems to have solved the puzzle.[122] He has postulated a complicated set of muscle contractions in which force is translated through the hyoid (tongue) apparatus. I think that the final answer still is not in place, but Lauder's account is the best so far.

CHAPTER 8

Physiology
and Behavior

It has the blood of a shark. . . .
—*G. E. Pickford*

$\mathbf{M}$odern study of the biology
of any living organism is sophisticated and complex. Even the
amateur naturalist, so valuable for basic field research, now
uses technologically advanced gear and collects comprehensive data—all a far cry from the old days when our main tools
were a pair of binoculars and a notebook and when "data"
consisted of lists of sightings. Happily, however, this is one
of the few areas in life where science has not completely ful-
filled its apocryphal mission of taking the fun out of everything.

The laboratory scientist nowadays uses a huge array of techniques, many of them depending on biochemistry, whether to study genetic relati nships or behavior. It is no wonder that with this wonderful armory of weapons, zoologists and paleontologists have been anxious to get their hands on fresh material of *Latimeria*. The preserved specimen in its tank of formalin is simply less useful than it was fifty years ago. But fresh material has been scarce, principally because no one has been able to catch a coelacanth "to order," but we have had to depend on the chance of being available when a fish has been brought still living to the beach by a fisherman. Now just as technology (principally submersible vehicles) begins to give us access to specimens, conservation considerations and the endangered status of the species may make us very cautious about killing *any* new specimens for research, and we may be thrown back to concentrating on work only needing observation.

Happily quite a lot has been accomplished with the material available, starting with the priceless Yale specimen in 1966 and then the two fresh specimens from 1972, plus some newer frozen materials. We are finally getting to grips with the study of *Latimeria*'s physiology and biochemistry, as well as its delicate soft tissues. We are even starting to work on behaviors other than swimming and feeding, behaviors, that is, that cannot readily be deduced from the skeletal anatomy or a preserved specimen.

BRAIN AND SENSES

The coelacanth brain has been studied quite extensively, principally in search of clues to the relationships of the group, the key question being, in comparison of the living forms: Are coelacanths closer to lungfishes or to amphibians? So far the results are inconclusive. *Latimeria* has a largish brain of relatively primitive type, but curiously, in mature fishes it may occupy only 1.5 percent of the braincase cavity,[123] the rest of the space being taken up with fats and oils. (This being the case, it may be hard to learn anything of the general shape of the brain from the shape of the inner surface of the endocranial cavity in fossils.) In the smallest *Latimeria,* of less than forty-five centimeters, the brain fills the cavity, so this is a growth phenomenon.

The brain has a well-developed region (optic tectum) for the processing of information from the eyes, and in general it resembles that of sharks and the lungfish *Neoceratodus,* but not the other lungfishes *(Protopterus* and *Lepidosiren),* which are generally reckoned to be the most derived or specialized of the lobe-finned fishes.

The sense organs of *Latimeria* are just what one would expect for fishes. The eyes are moderate in size, certainly not small, but not huge as in many really deep-sea fishes. We have noted some of the characteristics of the eye in Chapter 6. Several authors have mentioned that the eye of *Latimeria* is luminous or phosphorescent. But observations on the 1972 specimen and by Fricke clearly show that the eye is not truly luminous, in the sense of light-generating. The retina simply has a strongly reflective layer, or tapetum, and when a light is shone onto the eye, this gives back a bright reflection. The same thing can be seen with a domestic cat. The tapetum is an adaptation for vision in conditions of low light intensity.

The nose is well developed and not unusual. It has none of the specializations that one finds in amphibians; particularly it lacks the internal nostril—the passageway that in land vertebrates connects the nose to the throat region, allowing them to breathe air. This is the nasal organ of a fish rather than an amphibian.

The inner ear has large semicircular canals for balance as in other fishes and in birds, which also live in a three-dimensional world compared with us earthbound two-dimensional land animals. The lateral line system for detecting pressure changes and water movements, by which fishes detect the presence of other fishes, is also well developed, as one would expect.

There is, however, one unusual feature of the sensory organs of coelacanths that remains unexplained. Coelacanths have the usual eyes, ears, and nose, but in addition, there is the massive rostral sense organ, apparently unique to this group. It is found in the very first fossil forms and is one of the features that characterize coelacanths. Lying in the snout above the nasal organs, in front of the eyes, is a central sac with three pairs of external openings borne on short tubes leading from the chamber of the organ to the surface.[124]

The pattern of nerve connections linking the rostral organ to the brain shows it to be a sensory device. But what for? Is it for special chemoreception? This is unlikely, for the nose is already well developed. Depth or pressure reception? Unlikely, and in any case the lateral line is well developed to take care of this.

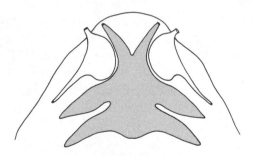

FIGURE 37   Position of the rostral organ between the nasal sacs in the snout of *Latimeria*. After Millot and Anthony.

In recent years a great deal of attention has been paid to the fact that most fishes (and some amphibians) have an extra sense absent in true land animals, electric receptors in the skin. It has been known since classical times that fishes like the electric catfishes, torpedo rays, and electric eels actually generate electric impulses, which they use as a defense, to stun prey, and sometimes even as a sort of radar to locate prey. One did not need to know what electricity was to recognize the jolt you got when stepping on an electric ray. The electric impulses, some very strong, are generated by special organs derived from modified muscle tissues.

It eventually became clear that even fishes that are not electric generators can detect weak electric disturbances in the water and have special sense organs for this purpose. One can even find evidence of electroreceptors in ancient fossil groups (including the rhipidistian and lungfish sisters of coelacanths).[125] The organs in question consist of minute chambers and canals in the superficial bones and tissues covering the head. The dimensions and distribution of the these organs are appropriate for electroreceptors, but of course, we cannot know for sure.

What kinds of electrical signals can there be in the sea for fishes to detect? The classic experiments done by A. J. Kalmijn of the Woods Hole Oceanographic Institution (now at Scripps) will give the answer.[126] Kalmijn studied sharks, whose snouts are well endowed with special electroreceptors called the ampullae of Lorenzini. These are arranged as a series of capsules containing the sense cells, connecting to the surface through tiny jelly-filled tubes. The jelly in the tubes is a specially good conductor of electricity. Sharks use this set of organs to detect prey that they cannot see or smell. A dogfish shark has the astonishing ability to find a flatfish buried in the sand simply by sensing the minute electric potentials caused by the firing of the muscles in the flatfish's gill as it sits quietly breathing under the sand. The signal is faint, to be sure, but salt water is a good electrical conductor.

It seems likely that electroreception is a primitive vertebrate feature, possibly related to the time when the first fishes lived only on the bottom and searched for food in the mud and detritus where eyes were not much good and noses might get confused.

It is now thought that the rostral organ of *Latimeria* may be an electroreceptor.[127] Physically the rostral organ is unlike the electroreceptor organs of other fishes, but it might, at a stretch, represent the coalescing of a large number of organs like the ampullae of Lorenzini of sharks.

There are two ways to settle the question: more detailed physical evidence and direct experimentation. Every kind of sensory cell has its own particular structure; electroreceptor cells in fishes are of a special type that could easily be identified in electron microscopic study. Specimens preserved in formalin or frozen for any length of time are no good for such techniques, so the critical observations have not yet been made.

As for experiment, Fricke observed a strange behavior in the coelacanths he filmed *in situ*. They occasionally did a lengthy "headstand," rotating to a vertical position with their noses down, just above the surface of the sand, and hovering in place. Could they be sensing something, searching for something, in this pose? With the nose or with the rostral organ? Fricke experimented very simply with an electric discharge into the water near the fish, and this frequently seemed to induce the headstand behavior.[128] The evidence is not yet conclusive, but an electroreceptive function for the rostral organ seems the best hypothesis at the moment.

ORGAN PHYSIOLOGY

The great significance of the Yale specimen, which was shipped fresh frozen, was that it opened up the possibility of learning something firsthand about how coelacanths work at the physiological level. Here I will mention three examples of the advances that have been made, examples that fit into the general pattern of revealing the biology of this fish as well as illuminate important questions about the evolution of vertebrates in general. There are many other observations, of everything from proteins in the brain to bile salts, that belong elsewhere than in this general work. Once again they involve finding the right hypothesis and then a way of testing it.

The story of *Latimeria* has always been, in good part, one in which women scientists have played a major role, starting with Marjorie Courtenay-Latimer. Grace Evelyn Pickford was a professor of biology at Yale when our specimen arrived in 1966; she had the office next to mine. Grace was a bit like J. L. B. Smith, gruff in manner and skinny in build, she was totally uninterested in matters of dress and appearance. She usually wore a man's shirt for convenience and blue jeans. For lectures she had an ancient blue serge suit, polished at the corners to a mirror finish. She lived for her work and her students. Having been trained in the older thorough traditions of natural science at Cambridge University and as a student of the blood chemistry and the physiology of fishes, Grace was as excited as I when our frozen specimen arrived at Yale in 1966.

On the great day (Memorial Day 1966) the fish thawed beautifully and with a perfect fresh smell. It was obviously well preserved. As the initial dissection proceeded, Grace collected blood and other body fluids, both in vials and dried onto slides. Then she took the vials up to her lab and was gone for a very long time.

Eventually Grace, looking shocked, came back down to where we were working and announced, "The blood is isotonic with seawater. It is full of nonprotein nitrogen. Keith, it is the blood of a shark." She glared at us, as if somehow it were our fault. Then she grinned and disappeared back upstairs to her lab, having made sure that there were plenty of liver samples for study as well. Pickford had made a tremendously important discovery.[129]

Fishes always face a running battle with the water in which they live. The blood (as well as other body fluids) of all animals is a complex tissue of water and dissolved salts, sugars and other chemicals, hormones, and trace elements with cells suspended in it. The precise concentrations of all the dissolved compounds are critical for the proper functioning of all the body's cells. The firing of a nerve impulse, for example, depends on a shift in chemical concentrations on either side of the nerve cell membrane. The concentrations inside and outside the cell must always be just right.

Not only must the concentrations of individual constituents in the blood be correct, but so must the total concentration of the body fluids. This obviously poses a problem for land-living animals that tend to lose water through evaporation and have to keep finding new water supplies, through drinking or from the food they consume. But the situation is no better for animals that live in water; in many ways the water is fishes' enemy.

The concentrations of basic blood chemicals turn out to be similar in all fishes, with some notable exceptions. Among the most important constituents of the blood are dissolved salts, such as sodium and potassium chlorides. In almost all fishes the sum of these amounts is roughly one-third the concentration of normal seawater. Obviously, therefore, fishes that live in freshwater have a blood that is much more concentrated (saline) than the outside medium, and marine fishes have a blood that is much more dilute than the sea in which they live.

For freshwater fishes the outside water is constantly forcing its way into the tissues through the process called osmosis, familiar from high school biology texts. This occurs principally across the gills, which afford a huge surface area. Fishes living in the seas, by contrast, have the opposite problem. As their blood is more dilute than the surrounding seawater, they tend to lose their body water by osmosis to the sea, again through the gills.

Only one sort of fishes, the hagfishes once again, the most primitive living fishes known, have basic blood salt levels at the same concentration levels as the ocean and thus avoid any osmotic problem.* Because hagfishes are generally so primitive, it is usually assumed that their blood chemistry represents the primitive condition for all fishes and that the reduced salt levels of tissue fluids in advanced fishes represent a special adaptation, a nuisance, as it were, that has to be put up with as part of the cost of sophisticated organ function and behavior

---

*Even so, of course, each of the particular constituents that make up the blood has a concentration that is totally different from the proportion of that chemical in seawater.

for which a lower blood concentration may be necessary.

There are, of course, rival explanations of the peculiar and seemingly contradictory blood chemistry and osmotic balance in modern fishes. The seas might have been more dilute in the Cambrian, when fishes arose, and thus the condition would be an archaic one preserved. This seems unlikely. Another possibility is that the early evolution of fishes occurred in freshwater, not the seas. In that case there might have been a very early reduction of blood salts in order to reduce the water intake problem, and once set in place, this could not be altered. Unfortunately each of these possible explanations is really only a plausible "story" (scientists use the pejorative term *scenario*)— very difficult to test. Little direct fossil evidence exists concerning the environment in which the first true fishes evolved, and it is hotly argued over.

In any case, all fishes except hagfishes are potentially faced with a constant problem of losing water or gaining water through osmosis. Oddly, for fishes the *seas* are a desert.

Freshwater fishes have to spend their lives pumping out the excess water taken in by osmosis. For this they use the kidneys to excrete a very dilute urine.* Fishes in the sea, on the other hand, need to pump out as little water in the urine as possible while eliminating their wastes.

Sea fishes, other than the hagfishes, have the constant problem of potential dehydration, and there have been two principal ways of avoiding it. The sharks and all their relatives solved the problem by adding to their blood concentration in an unusual way. The metabolism of all organisms produces a lot of waste nitrogen as a result of the breakdown of proteins. The result is ammonia, which is highly soluble, and fishes can easily flush it out of their tissues if there is enough water. That is easy in freshwater, but not in the seas. Ammonia is highly toxic and

---

*This in turn constantly exposes them to the danger of losing precious blood salts and sugars with the urine. So a special function of the kidneys in these fishes is not only to pump out water but also to act as superfilters and absorbing devices to conserve salts and sugars.

cannot be allowed to accumulate in the tissues. Sharks make a virtue of necessity and detoxify ammonia by combining it with another very simple molecule, carbon dioxide, to form urea. This process occurs in the liver and requires the operation of five enzymes in a special pathway. Urea is the excretory compound of choice for many organisms in which water is in short supply, including tetrapods such as us.

Sharks manufacture and then retain huge quantities of urea (and another compound called trimethylamine oxide or TMAO) in their blood at levels that would be lethal in humans. They retain enough that the blood acquires the same *total* osmotic concentration as the seawater, and osmotic gain or loss of water is eliminated. Indeed, by controlling the amount of urea and TMAO retained, these fishes can adapt to different concentrations of seawater. Some stingrays and some sharks can even migrate between rivers and the sea, and there are a few sharks, like the Lake Nicaragua shark *(Carcharinus leucas)*, that principally live in freshwater. (Sharks have a specialized rectal gland in the alimentary canal for elimination of excess salts.)

The other great group of sea fishes, the ray-finned bony fishes (Osteichthyes), has solved the problem of osmotic balance in a backward sort of way. These fishes actually drink seawater, retain the water, and excrete the excess salts through specialized cells in their gills. (Humans, like other tetrapods, cannot do this; that is why drinking seawater is fatal for people adrift at sea.) Most bony fishes do not have (or need) the capacity to synthesize urea in their livers.*

The question for us then became, What would be the case in coelacanths? The coelacanths, together with lungfishes and the fossil rhipidistians, are bony fishes. In theory *Latimeria* should drink in salt water and selectively excrete the salt. The blood chemistry should be like that of other marine os-

---

*A major problem for the bony fishes in the ocean is, of course, that actively secreting the salts back into seawater against a concentration gradient takes a lot of metabolic energy.

teichthyan fishes, like cod and tuna. But J. L. B. Smith was probably the first person to point out that with its primitive phylogenetic position, the osmoregulatory biology of *Latimeria* might offer interesting comparisons with sharks and bony fishes.[130]

All three genera of living lungfishes are strictly freshwater forms (although many fossil forms were marine) and, on the face of it, not much good for comparison. They do, however, have the capacity to synthesize urea in their liver, and they also have the capacity to tolerate large concentrations of urea in the blood. The reason is that lungfishes, or at least the African and South American genera, may spend many months of the dry season each year holed up in burrows. While aestivating, they depend upon protein as a source of metabolic energy, and this generates a lot of waste nitrogen and thus, potentially, a lot of toxic ammonia. So lungfishes convert the ammonia into urea in the classic way in the liver. This urea is then flushed out when circumstances allow. This sort of adaptation could have been another essential preadaptation when the fishy ancestors of the first amphibian tetrapod fishes started to stay out on land in order to be active, rather than merely aestivate and wait for the rainy season to return.

If lungfish can synthesize and accumulate urea, what would the situation be in coelacanths? Because of the close relation between coelacanths and lungfishes, Dr. George Brown had already speculated that coelacanths might, after all, be more like sharks than like other bony fishes.[131] Now Grace Pickford had shown that the blood of *Latimeria* was almost exactly like that of sharks and rays; they are the only marine bony fishes known to osmoregulate in this way.

It was a major discovery, and all that remained was to get Drs. George and Susan Brown to analyze the liver to see if *Latimeria* possessed the right constellation of enzymes for synthesizing urea. Within a few weeks they sent back the answer: Indeed, it did.[132]

Now there arose a new evolutionary problem. The previous notion that the two ways of osmoregulating were phylogenetically quite distinct (sharks osmoregulating one way and bony fishes the other) was obviously wrong. The condition in lung-fishes (also bony fishes) had been thought merely a secondary specialization for aestivation. That was wrong, too. Now urea synthesis and retention were shown to occur in both groups, and perhaps *it* was the primitive way of osmoregulation for all fishes, one that had been lost in those more advanced bony fishes (the ray-finned fishes) that lived in the ocean. But if so, why and how would those groups have given up that adapta-tion? Was it possible instead that urea synthesis and retention was an adaptation that had arisen not merely once but several times in parallel? The answer still is not known, but most ex-perts seem to favor the view that the adaptation has evolved more than once in the early history of fishes.[133] If so, the whole question of the environment of the first fishes, whether they arose in fresh or salt water, is once again reopened.

At least Professor Pickford's discovery that the blood of coelacanths is essentially iso-osmotic with seawater allows us to reject the theory (never very credible) that *Latimeria* needs to live in freshwater springs under the sea. Like freshwater sharks, coelacanths might perhaps be able to adapt to live under such conditions, but specimens caught so far have shown no sign of reduced blood concentrations. Recently Fricke has tested the salinity of the water in which the fishes he found were living. It was fully seawater. Even so, there remains one extra little puzzle. The blood of *Latimeria* is actually just a little *less* concentrated than seawater (about 95 percent of seawater). Coelacanths must use the seawater drinking device that other bony fishes use, at least to a small extent. They also have a rectal gland like sharks, possibly to eliminate excess salt. Why is their blood chemistry not more perfectly balanced with seawater? Do they need to use seawater as a source of essential salts?*

---

*When we first published the data from the Yale specimen, some people dis-puted them; maybe the fish had spoiled after all. When he took absolutely fresh blood and other tissues from the 1972 specimen before they had chance

DNA

Everyone nowadays has heard of the famous molecule DNA, the chemical whose structure encodes the information of the genes, of heredity and life itself. Each type of organism must have a different set of DNA molecules because each is genetically different. However, there is an interesting evolutionary puzzle: Different types of organism show huge differences in the *amount* of DNA found in their cells.

It has long been known that the most advanced ray-finned fishes have very small amounts of DNA in their cells compared with more primitive fishes. The numbers are roughly 0.8 picograms of DNA per cell in advanced teleosts like a puffer fish and 14.7 in a shark (a picogram is one-trillionth of a gram). Obviously the amount of DNA in the cells of advanced ray-fins is at least enough to carry out all the genetic functions. Do the primitive forms have more DNA than they need, or do they need more because of some different genetic function? Comparisons within tetrapods show the same situation: Some amphibians have very large amounts of DNA per cell, up to 205 picograms in the mud puppy *Necturus* and 192 in *Amphiuma,* while more advanced amphibians like the frog *Rana* and the midwife toad *Xenopus* have smaller amounts (15 and 6.3 picograms, respectively). The most advanced tetrapods—birds and mammals—have smaller DNA values (7 picograms in the laboratory mouse).[134]

There were several possible explanations of this. One is that DNA contents per cell were high in the ancestral vertebrates and that they have progressively decreased with evolutionary change in vertebrate lineages. The opposite could also be true. The forms that have very large amounts of DNA per cell may

to spoil, Bob Griffith could compare his results with those of Pickford and others who had used variously frozen and stored specimens. Those first data from the Yale specimen turned out to be remarkably good.

represent some kind of secondary aberration. Detailed analyses showed that forms with a lot of DNA per cell do not have lots of different *kinds* of DNA but merely different *numbers of copies* of the same elements.[135] Perhaps the amount of this "duplication" is a function of time: The older the lineage, the more chance the DNA has had to duplicate and increase in amount.

Obviously, to settle the question, it is important to know the ancestral situation, and here living fossil organisms come into their own. Surveys show that in primitive surviving *ray-finned* fishes the DNA contents per cell are in the shark range—10 to 15 picograms. However, the lungfishes have simply huge amounts of DNA per cell—160 picograms in *Neoceratodus,* 240 in *Lepidosiren,* and 284 in *Protopterus.* These are, in fact, the largest values known for any vertebrate.

To attempt to test more directly whether high or low values are primitive for vertebrates, my associates at Yale and I studied the size of bone cells in fossil lungfishes. We know, from comparisons within a wide range of living organisms, that in general there is a positive correlation between cell size and DNA content: Organisms with higher DNA values per cell have bigger cells. We could not measure the amount of DNA in the cells of fossils directly, of course, but it is fairly easy to measure size of bone cells, because the spaces occupied by the cells in the bone are often well preserved in fossil bone tissues such as scales. By comparing cell size among fossil taxa of different geological ages, we might at second hand be comparing, among other things, DNA content.[136]

This exercise turned out quite well. By comparing lungfish bone from different ages, we could show a progressive *increase* in cell size through lungfish evolution, thus suggesting that large DNA contents per cell are secondary. Cell size in the earliest Devonian lungfishes was compatible with a DNA content like that of sharks. Therefore, the primitive values for all fishes were probably low, but whether they were as low as in the puffer (0.8 picograms) we cannot tell.

Existing microscopic sections of fossil coelacanth bone showed a cell size closer to that of Devonian lungfish. Now, with coelacanth blood available, we had another opportunity to add some data. Would *Latimeria* have large, small, or moderate DNA values? The fresh blood samples taken directly from the 1972 specimens would tell us, and with the help of two colleagues an analysis was performed. Our prediction was that the value would be low (like a shark) rather than high. Previous attempts to estimate DNA contents from the size of nuclei in preserved body tissues had given provisional values in the range of 6 to 10 picograms.[137] In our new analysis we obtained a value of 13.2 picograms per cell, more than the lows for teleosts and mammals, far less than the highs in lungfishes and certain amphibians, and more in line with frogs and sharks.[138]

Coelacanth blood spoils extremely quickly after death, and the red cells break down. We had used samples of blood dried immediately after collection onto microscope slides. Another group of workers who used blood that had been stored refrigerated got a value of 7.2, which probably means that their sample had deteriorated, but the exercise could usefully be repeated.[139]

Another feature correlates with high DNA contents per cell in vertebrates. Organisms like lungfishes and a few amphibians with large cells and heavily duplicated DNA contents have very slow metabolic rates. The small amount of physiological evidence available (below) suggests that coelacanths also have a relatively low metabolic rate, and the DNA data therefore tend to confirm this.

All this raises the question, Why and how do DNA contents per cell in some advanced lines remain (or become) so low? One answer may be that a high rate of evolutionary change must involve a larger number of evolutionary transitions between otherwise stable populations and any excess duplicated DNA would tend to be eliminated during these transitions. In more conservative, slower evolving lines, there are fewer speciation events and fewer major evolutionary transitions. Therefore, there is less chance for "editing" out duplications, which instead accumulate.

METABOLISM AND RESPIRATION

Analysis of the body shape and swimming behavior predicts that *Latimeria* lives a rather quiet, even sluggish life. It drifts and hovers near the bottom, waiting for prey to come near, then darts after it in a short sprint. If this is true, we should find corroborating facts in the general metabolism and respiration of the animal. The first place to look is in the structure of the gill system. Professor George M. Hughes of Bristol University and his co-workers undertook detailed analysis of the gills of *Latimeria,* particularly using the two specimens collected by the 1972 expedition.[140]

These analyses show that *Latimeria* has gills that are similar to those of other fishes living at depths of two hundred meters or so and quite different from those living in shallow surface waters. The gills of *Latimeria* have a relatively low surface area compared with the body mass of the fish. The individual gill filaments are short, and there are few subdivisions into secondary filaments. They are the gills of a sluggish rather than very active fish and certainly not the gills of a fish that constantly cruises at speed in search of prey.

Studies of blood properties and tissue chemistry in *Latimeria* can also tell us a lot about the general metabolism of the animal, although not much work has yet been done. The blood, as everyone knows, is tremendously important to the metabolic function of vertebrates because in addition to carrying nutrients and hormones around the body and controlling disease through the white blood cells and the immune system, its principal function is to carry oxygen and carbon dioxide to and from the tissues.

The oxygen taken in at the gills is attached to hemoglobin molecules in the red blood cells (erythrocytes) and carried to the tissues and cells where it is needed. Carbon dioxide dissolves into the blood and is carried away from the cells where it is formed to the gills where it is discharged. In different organisms the blood has quite different capacities to carry these all-important respiratory gases. Much depends on how and where they live. Therefore, by studying the properties of the blood, particularly how much oxygen it can carry per unit volume, we can infer the sort of life *Latimeria* lives. We should be able to confirm whether it is a sedentary sort of animal or a constant fast-cruising fish.

The special properties are revealed most simply in analyses called oxygen dissociation curves, which measure the amount of oxygen that the particular kind of hemoglobin can bind as a function of the amount of oxygen available and different levels of blood acidity. Analyses performed on coelacanth blood show that *Latimeria* has a rather low oxygen carrying capacity; it would get tired very quickly under sustained exercise.[141] The optimum saturation of blood with oxygen occurs at temperatures between 15° and 20° C, which corresponds with depths below about 150 meters in the Comores. Further, it has very low levels of a crucial compound, cytochrome c, which is involved in releasing energy from compounds in which it is stored in the body and circulated in the blood. Once again, the morphological and physiological data come together with the limited sets of observations of live fishes.

The heart itself is unusually primitive in *Latimeria,* retaining in some respect the features of the embryonic stage of development of the heart in advanced vertebrates.

AGE

Latimeria reaches nearly six feet in length. How old must such a fish be? Most organisms leave some sign of their pattern of growth in the form of growth lines in the hard tissues. Trees do so extremely clearly. Almost every museum in the country has a section of redwood trunk with the growth rings marked to show the rings laid down in years like 1776 or even 1066. The skeletons of mammals also show marks, often in the teeth, although less clearly than do trees. Other skeletal tissues are less used because the hard tissues of mammals (except the teeth) are often physiologically reworked after they have been laid down. Wear patterns also often reveal age. For example, you can ascertain age in a horse up to eight years by the wear patterns on its teeth.

Easily cultured organisms like shellfish (oysters, for example), with their hard shells, are perfect organisms for experiments to test the accuracy and level of resolution of growth marks. In shellfish one can see not only yearly growth rings but marks of seasonal, monthly, and even daily fluctuations in growth. All these marks represent irregularities in the otherwise smoothly continuous laying down of new skeletal material during growth. Growth can be accelerated, slowed, or interrupted for many reasons. Winter lack of food, spring abundance of food, and the diversion of food resources to form eggs and sperm instead of hard tissue in the reproductive season, are the obvious ones. There are also sickness and injury and so on. In oysters one can sometimes even see a twice-daily mark for the tides, reflecting tidal variation in the abundance of food.

Fishes leave a variety of growth marks in their skeletons, particularly in the structure of their scales and the curious bones (otoliths—ear stones) that are found in the inner ear. Aging fishes on the basis of these growth rings has been taken to a refined art. Unfortunately all this works best in organisms that live in the highly seasonal environments of temperate latitudes, rather than in the tropics, where life is more smoothly continuous.

In common and easily reared organisms one can check raw observation against controlled experimental data from the laboratory. All this sort of control is lacking for coelacanths, but the scales, otoliths, and several head bones do have regular growth rings, similar to those of other tropical fishes. The markings are hard to see in some cases, and the numbers of "rings" may differ from bone to bone. Experts differ in the way in which the rings should be interpreted in *Latimeria*. One group believes that two rings are laid down each year as a result of interruptions of growth in January to February and August to September. In this case, the largest specimens known, which are females reaching about 180 centimeters in length, were eleven years old. The alternative view is that there is only one ring laid down per year, in which case the oldest specimens are twenty-two years old; this was Smith's first estimate.[142] In any case, these are relatively slow-growing and long-lived fishes, but not unusually so. By comparison carps have routinely survived in captivity for more than fifty years, and sturgeons more than a hundred years old have been caught by fishermen.

CHAPTER 9

Reproductive Biology

[W]e feel that there is a strong
possibility that the living coelacanth
is ovoviviparous.

—*R. W. Griffith and K. S. Thomson*

Hey, Jim, look what I've got.

—*C. L. Smith*

Somewhere below the surface of the Indian Ocean, along the steep sides of Grande Comore and Anjouan islands, lives at least one population of coelacanths. They swim, breathe, and feed in the currents that sweep along the island bases. We do not know how many fish there are. And although we know quite a lot about the species, until recently we had no idea how it reproduced.

A species can assure its survival only through reproduction. In a perfect world each individual organism would need to be responsible for the production of only one other over its whole lifetime in order to replace itself after death. Population size would then remain constant. If half the population were eaten per year by predators, however, then each surviving individual would need to be responsible for two new individuals on average every year. If not all the population were reproductively active or successful, then the remainder would have to do even more. And so on. No wonder, therefore, that most species spend their greatest efforts in garnering of resources for reproduction.

Latimeria's survival can hardly be considered secure. We are not sure of where it lives, how many there are, or what the rate of addition to the population(s) is through reproduction, or the loss through predation. We certainly do not know, although we should be concerned, what the effect of human predation (fishing) upon this species is. These facts of life make it critically important that we learn about the reproductive biology of *Latimeria*.

The diversity of ways by which fishes perpetuate their species far exceeds anything "invented" by the so-called higher vertebrates. It shows us that they have had a very long time to change and adapt to particular conditions. To be sure, many, perhaps most, fishes reproduce in the simplest possible way. The male and the female release eggs and sperm into the water, often with minimal courtship interaction. The egg and sperm find each other by chemical attraction and fuse, and the developing larva falls to the bottom or floats in currents, slowly developing into a hatchling. In this regard one always thinks of the single female cod that will lay as many as one million eggs or the herring that lays *only* fifty thousand eggs, all essentially abandoned to their own devices. This is apparently a sound strategy. By flooding the environment with huge numbers of offspring, the parents ensure that a significant number of young will survive without the parents having ever to bother about them. One can see in this all the principles of Darwinian natural selection at work. The fitter individuals— the largest and strongest, the faster and more efficient feeders, the fastest growers, and those that can avoid predators most readily—will have a better chance of surviving than the weaker. (I have to confess to a certain hesitation about this view, however. When one is dealing with massive predation on defenseless larvae floating in the plankton, pure chance may also play a huge role in "deciding" which ones survive.)

At the other extreme are more controlled strategies. In these cases, instead of producing huge numbers of young that are left to their own devices, the parents produce relatively few eggs and invest much more material, energy, and time in each. A female cod cannot give each of a million eggs a large quantity of yolk. Each larva is tiny and must feed for itself as soon as possible. If fewer eggs are produced, each one can be endowed with a large volume of yolk and can pass the important early stages of development independent from the vissicitudes of the environment. In addition, if there are few eggs, the parents can invest in their future in another way—by spending their time and energy in caring for the young until they are independent. Therefore, in many species of fishes we find elaborate nest-building behaviors, from the bubble nests of the Siamese fighting fishes to the nests of plant fragments made by sticklebacks. In some cichlids the parents may brood the developing fry in their mouths. The male sea horse carries the eggs safely in a pouch on his belly.

A prominent mode of parental investment in and care of the young is live bearing. We see this "solution" throughout the history of reproductive biology among the tetrapod vertebrates. Amphibians usually lay eggs into the water, although they are often attached to plants and protected by a "jelly" coat. But many species protect them in other ways, the midwife toad carrying the developing eggs around on its back rather like the sea horse. Reptiles and birds lay only a few eggs, each protected with a special shell and endowed with lots of yolk. Most birds guard the eggs in a nest and often care for the hatchlings as well until they can be fledged. In most mammals the eggs are never "laid," and the female's womb becomes the equivalent of the nest. The young develop inside the mother, totally protected and nourished, and are then born. After birth the young are further nourished by suckling. In some cases the young can care for themselves relatively soon after birth; deer and horses can run freely hours after birth, for example. In other cases, like humans, the young are helpless, and parental investment in the young continues for many years until survival is assured, both of the individual young and, thereby, of the species.

In live-bearing fishes one usually sees a particular strategy called ovoviviparity (egg-live-bearing), in which the female retains the eggs in her reproductive tract and protects them so that they grow up inside her body, being released at the "hatchling" stage. In many cases, apart from carrying the eggs around and protecting them in her special internal environment, the mother does relatively little more for the embryos. However, in some sharks the mother contributes actively to the nourishment of the embryo, and the condition approaches a true viviparity.

Ovoviviparity, like any form of live bearing, obviously requires internal fertilization by the male and an egg with a good yolk supply. The essential respiratory gases, oxygen and carbon dioxide, pass through the egg membranes. Certain food materials may in fact be available inside the mother through simple transfer from the uterine environment (absorption or swallowing of uterine secretions, for example), but basically the egg is self-supporting while it remains inside the mother.

The extreme in live bearing is the development of a special set of tissues that connects the developing embryo with the tissues of the mother, for nutrition and removal of waste materials. This is the adaptation that makes advanced mammals—the placental mammals—so successful. The placenta is a special set of tissues developed for embryonic support and nutrition, and it is shed as the afterbirth when it is no longer needed.

Not all mammals have the placenta, however. The ancient monotremes—the duck-billed platypus and the echidna, both from Australasia (more living fossils)—are mammals that lay eggs. All the marsupials (kangaroos and opossums, for example) give birth to highly immature young that are transferred to an external pouch to be hidden and nurtured for months at nipples. Very few fishes have anything like such an advanced reproductive biology, but in fact, some sharks show a surprisingly convergent pattern. In certain members of the families Carcharinidae—dogfishes and requiem sharks—and the Sphyrinidae—hammerheads—a placenta is developed in connection with the yolk sac. A sign of this pattern of development is that the eggs need be endowed with less yolk than if they were fully self-supporting.

There is one last and very curious variation on this set of themes of "viviparous" embryonic development in fishes. In a few sharks (for example, the mackerel shark *Lamna cornubica*) the embryos may feed cannibalistically on immature eggs or even on one another while within the uterus.[143]

What do the lungfishes, the closest living cousins of the *Latimeria,* do? Lungfishes make nests. All three genera scoop out depressions in the bottom where the water is shallow and lay large, yolky eggs. The male fertilizes the eggs in the nest by shedding its sperm onto the eggs as the female releases them. This requires a special courtship behavior to ensure perfect timing, for the sperm will not live for long independently in the water. In the African genus *Protopterus* the male guards the nest and may beat the water with his tail to scare away predators and perhaps to aerate it. Like amphibians, lungfish larvae have external gills, and in their reproductive behavior lungfishes generally act much like primitive amphibians, the cousins of both lungfishes and coelacanths.[144]

As for the coelacanths, up to 1975 the evidence was confused. As we have seen, there was quite good evidence that at least one fossil genus was a live-bearer. In these Jurassic fossils there is a smallish number of obviously highly immature individuals within the abdominal cavity, and they are positioned too far back to have been stomach contents. When Professor Watson described the specimens, there was general agreement that this coelacanth at least was a live-bearer, a not particularly surprising view considering that live bearing is reasonably common in all groups of fishes. The fossil evidence could never be conclusive, of course, but Watson's interpretation lasted for fifty years.

The ninth coelacanth to be caught (February 12, 1955) was a female with a swollen ovary containing "about ten ovocytes" (unshed eggs) of one to two centimeters in diameter. Number eighteen (January 1, 1960) was a very badly preserved specimen, but on dissection it had rather more fully developed eggs in the ovaries, one egg being seven centimeters in diameter, and was thought to be ripe—*"vraisemblement en cours de ponte."* These were the first coelacanth eggs to be seen. And for the first time the suggestion was raised that perhaps *Latimeria* was simply an egg-laying fish. Although there is no sign of a shell-forming gland in the female reproductive tract, eggs as big as seven centimeters should need protection. The detailed dissections of the French group showed nothing in any of the female specimens to suggest that they were live-bearing. Indeed, as there is also no obvious intromittent organ in the male, the possibility of live bearing seemed ruled out. *Latimeria* was therefore thought to be oviparous: *"nous avons pus nous assurer . . . que la reproduction est ovipare."*[145]

The question then lay largely unaddressed until 1972. As described in Chapter Four, members of the 1972 expedition by the Royal Society, National Geographic Society, National Academy of Sciences, and Muséum National d'Histoire Naturelle were lucky enough to be present when a large female specimen of *Latimeria* was brought in. This specimen proved to be gravid, containing nineteen huge eggs that had already been released from the ovaries. There was no evidence that development had been initiated. Further, the eggs had only the very thinnest of membranous coatings, and they were fluid to the touch. There was no sign of their being attached to the oviduct.

The more scientists involved thought about this, the stranger it seemed. The French had already shown that there were no shell glands associated with the oviduct and, therefore, no way for the female to protect the egg with a tough case. But how could such a large egg, the largest egg of any known fish, be shed into the water with so little protection? How could it survive?

Perhaps the eggs were normally laid into a nest, as is the case in lungfishes. Fertilization would occur by the male coming to shed its sperm on them. Then the delicate eggs must have been very closely guarded by one or both parents. In any case, the discovery of this heavily gravid female seemed to be evidence that the fossil specimens of *Undina* (=*Holophagus*) might have been mistakenly interpreted by Watson. Probably it was a case of cannibalism after all. *Latimeria* at least did not seem to be ovoviviparous.[146] Almost immediately thereafter Dr. Hans-Peter Schultze described some tiny fossil specimens of the Carboniferous coelacanth *Rhabdoderma* that were actually very immature young with yolk sacs attached. These specimens occurred in a significant size range and seemed to represent young hatchlings at the yolk sac stage. This seemed to be further evidence that the eggs were laid into a nest or the open water and that the young fry, with the yolk sacs attached, developed out in the open. Schultze stated: "This proves that the coelacanth is oviparous."[147]

But Dr. Robert Griffith, who had been a member of the 1972 expedition and was present at the dissection of that first gravid specimen and who was a student of Professor Pickford and myself, continued to worry about those eggs. He is a tall man, slow-speaking and quick-witted. Bob's idea of torture is having to wear a coat and tie. Periodically he would sit down in my office and say, "I don't believe it." Gradually he built up an argument that *Latimeria* must be ovoviviparous after all. He and I published this theory together in 1973, although I wish to emphasize that credit largely goes to Bob Griffith. It seemed crazy to many people, but it was an argument based on established comparative data, so we decided to stick our necks out.

The argument is as follows. Large eggs of the size found in *Latimeria* are found elsewhere only in fishes like sharks that are ovoviviparous. An egg that size is never laid without a shell. *Latimeria* is a ureotelic fish, meaning that it osmoregulates by a mechanism of urea synthesis and retention, as, again, do sharks. There is a special connection between ureotely and ovoviviparity in sharks, for the following reason. The specialized adaptation of urea retention may come fully into operation only rather late in development. Before it does, the young are constantly at risk of lethal osmotic imbalance because the outside seawater is more concentrated than the embryo blood. The only way to prevent this is to surround the egg in a closed, controllable environment. Sharks do this either by means of laying their eggs with the protection of tough egg cases (creating a closed mini-environment) or by means of ovoviviparity (in which case they grow up within the maternal environment). For example, the large six-gilled shark *Hexanchus,* which is found at the same depth as *Latimeria,* is ovoviviparous.

Therefore, if *Latimeria* is so much like a large shark in all other respects, especially ureotely, and does not produce a tough egg case, the fish should be ovoviviparous.[148]

In that case, how could fertilization occur? One possibility is that the structure of curious folds and tubercles around the external genitalia of the male actually constitutes some kind of erectile intromittent organ. But it also has to be pointed out that many vertebrates have internal fertilization without male intromission. Many amphibians fit this category. In the common European newt *Triturus,* for example, the male courts the female with elaborate wriggling movements and pheromone secretions, and she positions herself next to him. He deposits his sperm in a gelatinous package, and she slides alongside, places her cloaca over the package, and picks it up. The gelatinous material dissolves, and the sperm swim up into the oviduct. There seems no obvious reason why *Latimeria* could not do the same. But the trouble is the usual one: no direct evidence.

Quite inadvertently the puzzle was solved. Ichthyologists at the American Museum of Natural History in New York decided to dissect their big female specimen. This was specimen number twenty-six, acquired by the museum in 1962 (Chapter 2). The fish had a checkered history. It was caught off Anjouan on January 8, 1962, but the name of the fisherman was not recorded. Dr. Garrouste, the physician on Anjouan who was so instrumental in preserving the first French specimens caught in 1953, obtained the fish (presumably direct from the fisherman), bypassing the governmental channels. He first offered to sell the fish to J. L. B. Smith in South Africa. Smith decided that he did not want it but knew that the American Museum of Natural History would. Dr. Bobb Schaeffer was then head of the Department of Vertebrate Paleontology at the museum and very much wanted the specimen. At first it was difficult to arrange the transfer because the French authorities did not want Dr. Garrouste to sell it. Finally they released the fish in exchange for a thousand dollars of medical supplies for Garrouste, donated by a patron of the museum.

In 1975 Dr. Charles Rand, a hematologist on the faculty of Long Island University, wanted to get tissue samples from the spleen of a coelacanth for comparative purposes. Since so many other coelacanth specimens had by that time been dissected and used for research and, indeed, other superficial dissections of this specimen had already been made at the museum, the old injunction that it could be used only for display was obviously inoperative. When Dr. C. L. Smith from the department of ichthyology at the museum and Dr. Rand opened up number twenty-six, to their amazement, they found five almost fully developed young *Latimeria* inside the swollen oviduct, each with the remnants of a large yolk sac still attached.[149] Each was nearly thirty centimeters long, and the yolk sac still measured six centimeters. *Latimeria* is indeed a live-bearer, so Professor Watson must have been right about the fossils after all. We do not know at what size the young are

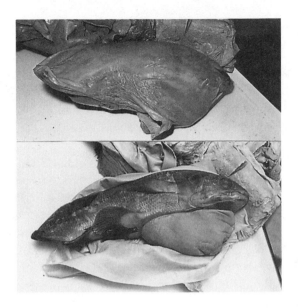

FIGURE 38   The swollen uterus of specimen number sixty-two contained five thirty-centimeter embryos like this. COURTESY OF AMERICAN
MUSEUM OF NATURAL HISTORY, PHOTOGRAPHIC LIBRARY, NEGATIVES
66635 AND 66637

born and whether they are born with any remnant of the yolk
sac still attached or whether it has all been resorbed by that
point. A forty-three-centimeter specimen, presumably free-
swimming since it was caught by hook and line, that was caught
in 1973 showed no sign of a yolk sac, so all we know is that they
are borne at a size somewhere between thirty and forty-three
centimeters.

The Carboniferous fossils that Schultze had described as
free-living yolk sac larvae were from a genus (*Rhabdoderma*)
that may have lived in brackish waters rather than the sea. In
this case (if the water was less saline than the body tissues) it is
possible that ovoviviparity was not needed, but economy of
hypothesis suggests the strong likelihood that this genus was
ovoviviparous also. When our 1969 expedition caught the
shark *Hexanchus* in the western Indian Ocean, we would occa-

sionally take a gravid female. The highly stressed fish would shed its embryos onto the deck when it was heaved on board. Perhaps these fossil embryos of *Rhabdoderma* were also shed by stressed females, dying in the swamps of ancient coal deposits.

Of the five coelacanth babies discovered at the American Museum of Natural History, one was given to the British Museum (Natural History), and one was given to the Muséum National d'Histoire Naturelle in Paris in exchange for an adult specimen (number twenty-five, a male). One specimen was sent to the Children's Hospital of San Francisco to be specially serial-sectioned, which means that it would exist in the form of a complete series of microscope slides, each representing a slice a few thousandths of a millimeter thick. This series of slides would enable students to see the finest details of anatomy at any point in the body and to reconstruct, for example, the finest details of nerves and blood vessels passing through the organs. (This project has unfortunately stalled and is incomplete.) One of the specimens remaining at the American Museum has been prepared by a special method that renders the soft tissues transparent and reveals minutiae of the internal skeleton. The last specimen has been preserved intact.

FIGURE 39   Each of the five embryos had its yolk sac (remnant of the egg) still attached, just like the fossil embryos from much smaller species described earlier (lower right). After Smith, Rand, Schaeffer, Atz, and Schultze.

At last the puzzle of reproduction in *Latimeria* has been solved—and another hypothesis based on reasoning from indirect data has come out right.[150] Recently Dr. James Atz of the American Museum and Dr. John Wourms of Clemson University have added a final new wrinkle to the story by suggesting that *Latimeria* might in fact show the odd "cannibalistic" feeding of the young on unfertilized eggs, as in sharks like *Lamna*.[151]

For all our euphoria at confirming what seemed at the time to be a daring hypothesis, in one very important respect all this new information about the mode of reproduction in *Latimeria* is alarming. Because the gestation period for *Latimeria* must be quite long, perhaps as long as a year, the rate of renewal of a small population would necessarily be slow. And because the female is single-handedly responsible for the well-being of a small number of live young, rather than huge numbers of eggs, any time a female is caught there is potential for significant reduction of the capacity of the population for renewal. We may have discovered an important vulnerability of the living coelacanth.

Coelacanth Relationships and the Origin of Tetrapods

> [T]here was then a long-held belief
> that coelacanths were close to the
> ancestry of tetrapods.
>
> —*P. L. Forey*

The new information that we have gathered on the anatomy and physiology of *Latimeria,* and the insights this gives us into the biology of fossil lobe-finned fishes, allow us to reexamine the relationship of coelacanths to other groups of fish. This is important because just when things ought to have been getting simpler, some rather offbeat ideas have been introduced, and the situation has become unnecessarily confused, confirming once again the truth of Alexander Pope's adage "A little learning is a dangerous thing."

In fact, the analysis of relationships among groups of organisms (and perhaps a lot of other science, too) is a subject in which everyone tends to have a strong point of view. In all the mystery and glamour surrounding a living fossil like *Latimeria*—the romance of its distant island home and the drama of efforts to study it—there has been a lot of exaggeration and hyperbole. Because one of the principal fascinations of *Latimeria chalumnae* has been its possible genetic relationship to man's distant ancestors among the other lobe-finned fishes and to the first amphibians of the Devonian, it is particularly important to clear up two key issues: To what group or groups are coelacanths most closely related, and what light does *Latimeria* shed upon the question of the origin of land vertebrates?

Ever since it was posed at a meeting of zoologists in 1978, the following conundrum has served as a sort of shibboleth among scientists who are concerned with the study of systematics—the genetic relationships among different organisms. The question is this: Of the following three organisms, which two are the more closely related, a lungfish, a salmon, and a cow? It sounds silly, but it is serious, and the correct solution to this conundrum helps us work out the question about the significance of *Latimeria.*

From the very beginning (which means March 1939) one of the questions driving public interest in *Latimeria* has been the question of its relationship to the tetrapods—amphibians, reptiles, birds, and mammals. As we noted in Chapter 3, zoologists have long been certain that it was from *some* member of the lobe-finned fishes (Sarcopterygii) that the first Devonian land vertebrate arose, in the form of a fishy-looking creature that breathed air. But it has also been clear for at least a hundred years that of the three or four known lobe-fin groups, it is

unlikely that the coelacanth group itself contained that ancestor. Rather, the ancestor must have been either a member of
the "osteolepiform rhipidistians" or a "lungfish." Neverthe-
less, and doubtless because the hint of a special relationship to
higher vertebrates and therefore to man carries enormous
popular appeal, *Latimeria* is often referred to as a missing link
between fishes and man. Sometimes this occurs as an elision, a
sort of shorthand forced by accident or press of space (I have
been trapped this way myself). Sometimes the hint is made
deliberately to court attention, for scientists crave attention no
less than other mortals and have the extra problem that this
attention is often not unconnected to the supply of research
monies.

As study of *Latimeria* has proceeded, all sorts of information
have been collected that could never have been gained directly
from fossils—the structure of the soft tissues, for example, and
their physiological function and behavior. With this has come
a fascinating new puzzle. While coelacanths seem obviously to
be related, if only indirectly, to the ancestors of the tetrapods,
in certain other respects (blood chemistry and urea retention,
for example) they look very much like relatives of the *sharks.*
But sharks and their kin (the skates and rays and the chimaeras
or ratfishes) are usually thought to be totally removed in any
phylogenetic sense from the bony fishes (Osteichthyes), the
big group to which all the lobe-finned fishes belong. The puzzle of what this all means is more likely to appeal to other
zoologists than to the general public, but it is just as fascinating as the question of tetrapod origins.

So we need to review the state of science concerning these
two problems and to see what light *Latimeria* actually does
shed on the question of tetrapod origins.

At a now-famous meeting of vertebrate zoologists in Reading, England, in 1978, the zoological conundrum about the
lungfish, the salmon, and the cow caused a mild sensation. In
retrospect it should not have caused more than a blink of the

eye. The most obvious answer is, of course, that the two fishes *appear* to be more closely related than either is to a cow. They both are fishes after all. But . . . blink the eye, and it is evident that the lungfish and the cow are more closely *related* (although physically less *similar*) because they belong on the closer line of descent. The lungfish is a lobe-fin and, therefore, at least a sort of cousin of the tetrapods (including cows), while the salmon is merely a cousin of *all* that assemblage. An analogy would be the relationships among the late Prime Minister Indira Gandhi of India, Queen Elizabeth II, and Kaiser Wilhelm II (ruler of Germany during World War I). The two ladies are late-twentieth-century female rulers of countries, but the queen and the old kaiser both are descendants of Queen Victoria.

To put the answer to the question in another way, the salmon and the lungfish share many similarities, all related to the fact that they are fishes. Both have fins and gills, for example. But these are primitive characteristics that are common to all fishes, rather than unique connections between lungfish and salmon. While lungfishes and cows share fewer characteristics missing in salmon, those they have are all advanced features found only in a single lineage (lobe-fins and their descendants) and nowhere else. The point is that in deciphering relationships among organisms, one must look for uniquely shared characteristics. Two or more groups can only share a specialized (derived) character, except by descent from a common ancestor, if the character has evolved more than once by convergence or parallelism. (This is a tricky point in practice. Some workers think that parallelism is extremely rare; others think it may be common.)

These simple principles by which to analyze relationships among organisms have already been used at several points in this story. Now we have to apply this principle to the relationships of coelacanths.

The first problem is, What characters shall we analyze—anatomical, physiological, behavioral? The second is, What is our frame of reference? We need some standard against which to calibrate whether something is primitive, advanced, or convergent/parallel; we need one that is acknowledged to be very primitive. For this, we can take a primitive ray-finned fish, such as a sturgeon, the bowfin *Amia,* or a fossil Devonian ray-fin, and of course, an amphibian or a reptile is our point of comparison at the other end. If we compare ray-fins, *Latimeria,* living lungfishes, and any amphibian, it always turns out that lungfishes are the closest relatives of tetrapods, and the more we learn about lungfishes and *Latimeria,* especially details of their soft anatomy and physiology, the firmer the relationship seems to be. However, while this exercise seems very sound, we should also include the *fossil* lobe-fins.

Unfortunately the number of characters available to analyze fossil groups is smaller. Instead of our analysis being heavily weighted toward the analysis of characters like the soft anatomy of the brain, blood system, musculature, and visceral organs, it obviously must rely almost exclusively upon skeletal features. But the results change.

When we add to the analysis from one species of fossil osteolepiform rhipidistian fish (the Late Devonian species *Eusthenopteron foordi* from eastern Canada, for example) and the first fossil amphibians from the Late Devonian of Greenland (*Ichthyostega* and *Acanthostega*), a wholly different result appears. Now the fossil osteolepiform and the amphibians are most closely related, lungfishes are the next most closely related, and coelacanths are the most distant group.*

The skeletal features crucial in the analysis are in the head and lobed fins. The pattern of bones in the dermal skull (that is to say, the external bony covering of the head) is point for

---

*This being the case, one must conclude that the soft anatomy features held in common between lungfishes and tetrapods must also be characters common to the osteolepiforms. Unfortunately this will be impossible to test directly, and naturally one is a little uneasy with a result that depends so much on adding or subtracting one taxon and with a process that shifts from one set of evidence to another.

point the same in the osteolepiform *Eusthenopteron* and the earliest amphibians. It is quite different from that of lungfishes. But even more remarkable is the similarity of the paired limbs of osteolepiforms and tetrapods, especially considering their very different functions on land and in the water (see later in this chapter).

Coelacanths and lungfishes have essentially the same skeletal structure in the lobed fins, with a strong central axis of bones in a single row, with fin rays branching off from the sides. The osteolepiform Rhipidistia, on the other hand, are unique in having a branched internal fin skeleton in their paired fins (not the second dorsal or anal). In the pectoral and pelvic fin skeleton, the first or proximal element was a single bone. This articulated more distally with a pair of bones. One of these branched again into two, and then these branched again. The fin rays were then borne at the end of this fan of internal elements.[152]

It is easy to see here exact homologues of the tetrapod arm or leg skeleton (including our own). The proximal element is the humerus (upper arm) or femur (upper leg), followed by the ulna and radius (forearm), and the tibia and fibula (lower leg). Next come the bones of the wrist and ankle, borne on the ulna (in the fore limb) and tibia (hind limb) respectively. Distal to this, the precise homologies break down; naturally enough, one cannot trace a complete homology of the more distal bones of the fingers and toes (which are strictly a tetrapod invention) in the fin skeleton of a Devonian fish. But the similarities in the proximal parts of the limb are remarkable and too complete to be a coincidence or due to the retention of some common primitive condition that every other group has lost.

One might think that the tetrapod limb looks a bit more symmetrical than the analysis just given; the two bones of the lower limb (ulna-radius and tibia-fibula) both appear to branch into the hand or foot. However, if we look at the sequence of development of the bones in any tetrapod embryo, we see

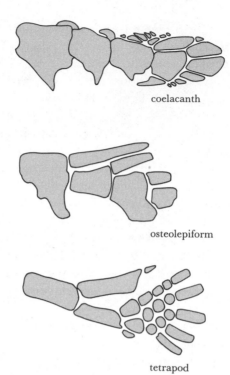

coelacanth

osteolepiform

tetrapod

FIGURE 40   Internal skeleton of the left pectoral fin in *Latimeria* and the Devonian osteolepiform *Sterropterygion,* compared with a hypothetical scheme of the tetrapod limb. Recent evidence suggests that the limb of the very earliest amphibians had more than five digits in the hand and foot.

clearly that the radius and fibula do not branch; they are connected just as in the fossil osteolepiform rhipidistian.*

All this sort of analysis comes under the heading of frustrat-

*There are many more points of similarity between osteolepiforms and tetrapods that are unique to this pairing. The humerus is twisted along its axis in both, for example. On the other hand, an embryo of an Australian lungfish was recently discovered with the first element of the fin axis paired rather than single. This at least suggests how the osteolepiform-tetrapod condition could have arisen from the coelacanth-dipnoan condition.

ing work, but if it were all clear-cut and simple, there would be no intellectual pleasure in it.

Adding to the frustration (or the fun, depending on your point of view) is the new problem of whether coelacanths are related to sharks. Have zoologists been missing the point? Perhaps coelacanths are not lobe-finned fishes at all. In other words, if we had added a shark or two to the group of organisms just analyzed, we would come up with a very different answer.[153]

The "shark relationship" argument arises because in certain soft tissues and physiology, the living coelacanth does indeed share features with sharks: retention of urea in the blood for osmoregulation, for example.*

Only a few workers strongly support a coelacanth-shark relationship, but they have defended it very hotly. One can readily see why someone would pursue the "shark hypothesis," even against heavy odds. To overturn a hundred years of opinion and establish a new, opposite position is every scientist's dream. Happily, help is at hand. All our suites of morphological and physiological characters have their basis in genetics. Most people agree, therefore, that if we could compare directly the characteristics of the genetic code—the detailed sequences of structure along the DNA molecules in the chromosomes and their related RNA molecules—we would have a much cleaner set of data to work with. There would still be the possibility of convergence and parallelism, to be sure, but molecular systematics is obviously a very promising tool, even though you cannot use it on fossils.

Results of molecular analyses, based on tissue samples taken from the 1972 specimens and subsequent frozen material, are just starting to come in. They offer no support at all for the shark theory. Sharks and coelacanths are very distant

*Other features trotted out in favor of a special coelacanth-shark connection include the nature of the pituitary complex of endocrine organs in the brain, the presence of a rectal gland associated with salt excretion, high levels of trimethylamine oxide in the blood, and similarities in the islets of Langerhans in the pancreas.

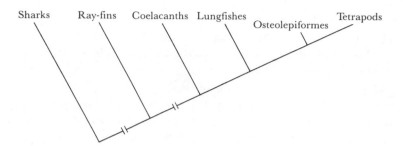

FIGURE 41 Relationships of the lobe-finned fishes, tetrapods, ray-fins, and sharks.

from each other. Therefore, any similarities one can find in their soft tissues must be shared primitive conditions or due to secondary parallelism. The shark theory can safely be put to rest.

Predictably, using the same molecular techniques to compare lungfishes, *Latimeria,* and living amphibians so far has produced no useful resolution of the problem of interrelations. The lobe-fins and tetrapods are clearly related, more closely than any of them is to ray-finned fishes or to sharks. But rival phylogenetic schemes *within* the group find almost equal support in the data, probably because we are dealing with taxa that diverged and followed separate paths long ago.

Because of the fact that fossil organisms can never be analyzed in terms of soft tissues or molecular characters, the relationships among lungfishes, rhipidistians, and primitive amphibians may never be fully resolved to everyone's satisfaction. However, if we all agreed, there would be nothing left to research and write about. But there does seem to be consensus about the position of coelacanths.

The Origin of Tetrapods

Even if *Latimeria* and all the fossil coelacanths are not close to the immediate ancestors of tetrapods, this living fossil fish has shed much light upon the fascinating questions of how, why, and where the first land animals evolved.

While careful scientists have avoided describing *Latimeria* as the missing link between fishes and tetrapods in a direct phylogenetic sense, we all have hoped that study of the structure, physiology, and behavior of this living fossil would illuminate our studies of the Devonian origin of tetrapods from other lobe-fins. In some respects *Latimeria,* a marine fish, has proved less valuable for such comparisons than the living freshwater lungfishes have been, especially when it comes to respiration and metabolism. In other respects, however—for example, as the only fish with an intracranial joint—*Latimeria* has been very important.

The oldest fossil amphibian we know is Late Devonian in age. In the uppermost Devonian strata of East Greenland there are at least two and probably three different genera of amphibians: the Ichthyostegalia, of which the best-known genus is *Ichthyostega* (fish-plate). There are amphibian trackways from the Upper Devonian of Australia and also a highly dubious lower jaw.[154] Therefore, we can probably bracket the date of origin of the Amphibia as no later than the Late Devonian and probably no earlier than the beginning of that period (probably not as early as the Middle Devonian).

At that time the lobe-finned fishes were diversifying widely but still modestly. By conservative estimates we know twenty-five to thirty genera of lungfishes and the same number of genera of osteolepiform fishes from the Devonian. The coelacanths numbered only fifteen genera or so. The osteolepiforms and lungfishes had already invaded freshwaters and were found in a wide range of environments, from the purely marine environments of coastal reefs and lagoons to the brackish water of large river estuaries to freshwater ponds, lakes, and streams. The rhipidistian and dipnoan lobe-fins were the dominant predatory group of these environments, taking fish, arthropods, mollusks, and probably any kind of animal prey that they found in the water. There were two distinct patterns of feeding mechanism. The lungfishes had developed a crushing feeding mechanism, using massive tooth plates to grind up hard-bodied food. Most of the other lobe-fins retained an intracranial joint and rows of sharp teeth for taking prey whole. The lungfishes were bottom-living fishes cruising along looking for arthropods and shellfishes. The other lobe-fins fed primarily on fishes although they no doubt took arthropods and soft-bodied invertebrates when they found them.

FIGURE 42 Restoration of the most completely known Upper Devonian amphibian, *Ichthyostega,* from East Greenland. After Jarvik.

At this time all the lobe-fins probably had functional air-breathing lungs. At least we can be reasonably sure that the lungfishes and osteolepiforms did. Lungs arose as paired outpocketings of the anterior gut, in the throat region, that were used for auxiliary oxygen uptake. The argument is usually made that air breathing gives a great advantage in shallow tropical freshwater environments. The solubility of oxygen in freshwater is low, and lower still in warm water. There is at least twenty times as much oxygen in a given volume of air than there is of water. Therefore, if oxygen were in short supply in the water, air breathing would be an advantage. Oxygen is often in short supply in tropical freshwater pools and swamps because the high temperature produces a high rate of bacterial decay of plant and other organic matter; the oxygen gets used up by the bacteria, and the other organisms in the water asphyxiate.[155]

It is often assumed that lungs arose in freshwater because modern fishes with lungs are primarily freshwater forms. Marine waters, with their constant wave action, are thought always to be better aerated. But in fact, the solubility of oxygen is even less in warm salt water than in warm freshwater and may become very low in tropical marine pools and lagoons. It is by no means certain that lungs did not arise in the seas after all.[156] Be this as it may, there is no doubt that lungs would have been an advantage to all lobe-fins living in shallow waters, and the arrangements of the great continental landmasses (and their coastal zones) in the Devonian was such that the majority of the environments (salt water or fresh) in which lobe-finned fishes lived were tropical and therefore hot.

Tropical environments are often very productive biologically, but tropical aquatic systems can be heavily stressed by fluctuating rainfall and temperatures. In just these conditions the early Devonian lobe-fins first invaded freshwater systems from the sea, and because of the successful previous invasions by plants, they found there an abundance of food and favorable environments. In the Devonian, for the first time, there were full freshwater, semiterrestrial, and even (in the lowlands at least) terrestrial floras, and these in turn supported new faunas of worms, mollusks, and arthropods. We like to point today to the importance of coastal and inland wetlands—swamps, marshes, and pools—as components of complex ecosystems. These same environments must have been even more important in the Devonian.

Considering that freshwater and estuarine and coastal environments in the Devonian must have been highly stressed, why did fishes invade the even more precarious terrestrial realm? They certainly did not do it all at once. Instead, they secured a toehold (using their newly evolved tetrapod feet, of course) in semiterrestrial environments—marshes, for example—and in the cover of permanent vegetation along the fringes of waterways. They continued to breed in water, but by being indepen-

dent of the water for at least part of their daily lives, they could
survive better than ordinary water-bound fishes. They could
escape predation in the crowded pools of drying up river-
courses by squirming through the mud from a shrinking pool
to a fresh one. They could tap new food sources in among the
roots of emergent vegetation. They could escape the parching
heat under emergent vegetation. They could lie in wait for
prey living in the water, seize it, and bring it onto the bank to
eat safely. They could lay their eggs and guard them in isolated
pools and swamps away from fishy predators.

But how did fishes manage all this physically? The answer is
that the necessary adaptations arose not on land but in the
very waters that they were now abandoning. *Ichthyostega,* although it is clearly an amphibian in its limbs and in lacking
gills, was nonetheless fishy in appearance. It retained a long
tail for swimming, complete with fin rays. The two key adaptations—air breathing and the tetrapod limb—probably arose
first as adaptations for improving life in the water. Only later
did lungs and limbs turn out to be even more useful on land.
Air breathing arose, as we have already noted, as a system of
auxiliary respiration when oxygen in the water was in short
supply. The fish needed water, however, to get rid of carbon
dioxide; it always required a wet skin.

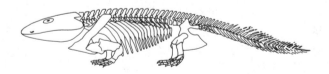

FIGURE 43   The skeleton of *Ichthyostega.* After Jarvik.

The tetrapod limb started out from the basis of the lobed
fin. As we have seen, this is a fin with a stout internal muscular
and skeletal axis, to which the fin rays are attached. Such a fin
has good mobility and first was used for complex swimming
motions, particularly slow swimming as the fish hovered

around, waiting for prey, perhaps very much as *Latimeria* does, or as *Neoceratodus* looks for prey on the bottom. In shallow water the lobed type of fin was useful also in pushing off against the bottom to help in locomotion. Elongate lobe-finned fishes could squirm along using the body the way an eel does on land, but short or stiff-bodied forms would have needed the paired fins as well.

The major use of the paired fins at first may, however, not have been in locomotion so much as in connection with respiration. In deep water, in order to take a breath of air, a fish like a lungfish can swim upward to the surface and hang there almost vertically.[157] However, in shallow water the front end of the body has to be flexed and propped up out of the water. If the water is shallow enough that the fish is partly out of the water, it starts to become heavy, and then the fin's propping-up function of support is even more important because it prevents the weight of the unsupported body from bearing down on the lungs and hampering ventilation.

FIGURE 44   The first prototetrapods used the fore and hind limbs differently. A prime function of the "arms" was to bear the weight of the head and trunk.

Right from the first stages of the transition to a terrestrial function of the paired limbs, the fore and hind limbs will therefore have had different functions. The fore limbs served principally for support and for propping the fish up. The hind limbs acted as levers in lateral body movements, devices for pushing backward against the bottom to assist in locomotion. The fore limbs were only partially functional in locomotion at first, and the difference between the fore and hind limbs is permanently recorded in the opposite flexure of the elbow and

knee joints. The elbow joint is flexed backward so that the
"hand" is brought forward under the "chest" to bear weight
under the anterior trunk. The knee is folded forward. (This
differential in function persists in all tetrapods in the sense
that the fore limbs bear more weight than the hind limbs because of the fact that the fore limbs have to support the heavy
head and neck. You can test this by putting a dog on two
scales. The weight recorded by the scale for the fore feet will
be some 50 percent greater than the weight borne by the hind
feet.)

In tetrapod locomotion the two sets of paired limbs work
differently from each other. The fore limbs reach forward and
pull backward against the ground. The hind limbs are drawn
up, flexing at the knee, and then *push*. This again neatly reflects the original difference in function between fore and hind
fins in aiding respiration and locomotion, respectively, at the
fish-tetrapod transition.

In comparing the skulls of the first amphibians and the
osteolepiform lobe-finned fishes, one notices immediately that
the main dermal bone patterns are identical. The difference is
that the fish has the opercular and gill apparatus attached to
the head and *Ichthyostega* does not. But while the arrangement
of bones is similar, the proportions of the head are different
and all the amphibians have lost the intracranial joint. It seems
to be the case that the fish-tetrapod transition was marked by a
relative lengthening of the anterior portion of the head (in
front of the intracranial joint) so that the head became more
crocodilelike. This is a typical adaptation of a fish-eating pred-
ator and perhaps indicates that the lobe-finned osteolepiforms
were more omnivorous and that the transition to amphibian
life involved a specialization of the diet for piscivory. A simple
geometrical analysis of the intracranial joint shows that it is
possible to change the proportions of the skull only within
certain limits before the joint mechanism becomes unwork-
able. No osteolepiform seems to have passed that limit, but the
first amphibians did. A remnant of the joint, in the form of a
suture across the braincase in just the right place, is found in

Ichthyostega, thus confirming its lobe-finned ancestry.

When the first amphibians came more fully out of the water, weight became a real problem; the backbone changed from being a compression member, preventing the body from fore-shortening when the swimming muscles contracted, to a girder from which the viscera were suspended. The backbone and the rib cage became extremely important parts of the skeleton. In *Ichthyostega* the ribs are very large and partially overlapping, making a solid box that would have protected the lungs from collapse under the weight of the body. In later tetrapods that could bear the weight of the body up on the limbs, the rib cage is lighter. It is unlikely that such a massive rib cage could have been expanded and contracted for breathing movements, so how were the lungs ventilated? They may have done it by force-pumping air down into the lungs by the mouth (as modern frogs do) or by using the liver and other viscera as the plunger in a pump, moving them up and down the abdominal cavity (as alligators do).

While respiration was not a major problem for the fish-tetrapod transition but simply required a transfer from a dual

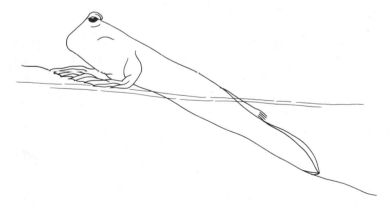

FIGURE 45 The tropical mudskipper *Periophthalmus* is an advanced ray-finned fish, whose "legs" are formed from modified pelvic fins.

system (air-water) to air breathing alone, water loss must have been a constant trial for the first tetrapods. In all probability they did not stray far from water, and certainly they must have bred in water and cared for the eggs and young there. But hot tropical environments must have presented a constant risk of desiccation unless the first amphibians always stayed under the cover of moist vegetation. Even then, both carbon dioxide elimination and ammonia will have been a problem. Once again the (pre)adaptations of fishes may have come in handy. We have noted that the lungfishes have the capacity to synthesize urea in their livers in order to immobilize ammonia and conserve water. All lobe-fins used it as a device allowing them to survive in salt water as well as fresh. We have to conclude that the first tetrapods had this capacity also and had inherited it from their lobe-finned forebears. In fact, there must have been a neat dichotomy here. The first amphibians used the adaptation of ureotely to exploit terrestrial conditions, while the lungfishes used it to avoid them. The lungfishes used ureotely to hide in the mud and wait for the water to return.

This then allows us to ask, Did the first amphibians live only in freshwater environments? Or did they also invade the rich coastal lagoons and marshes of the Upper Devonian? We do not know the answer. The few amphibian fossils found so far came from freshwater deposits, but the organisms may well have had the physiological adaptations to survive in salt water as well as fresh. But unless amphibians evolved twice in parallel, they must have had some sort of marine tolerance because the two places where fossil amphibians have been found in the Devonian—Greenland and Australia (if the identifications are correct)—were separated by areas of ocean during the Devonian.

Population Size Conservation, and the Future of *Latimeria*

... senseless slaughter.

—*J. L. B. Smith*

Science can be maddening because one never gets information in the right order. One often learns the most esoteric details before the most basic. Sometimes this skews a subject out of proportion. For example, radioactivity and X rays were studied quite casually for a long time before it dawned that they were extremely dangerous. The same is true of chlorofluorocarbons, while nitroglycerin was a wonderful explosive long before it was discovered to be useful as a heart drug.

In the case of *Latimeria chalumnae,* scientists have been acquiring specimens for study, by means of bounties offered to the Comoran fishermen, without anyone having the foggiest idea of how many fish there are lying below the surface of the Indian Ocean. If there are hundreds of thousands of individuals, of course, catching a few each year will not matter. If there are only hundreds, then "a few" becomes a very dangerous number. Of all the reasons for wishing to know what the population size is, the most urgent is the need to know whether we are driving the fish to extinction by the current fishing pressure for specimens.

POPULATION SIZE

No one thinks there are large numbers of *Latimeria.* A living fossil species like this is expected to be rare, especially because its distribution seems so restricted. Smith always guessed there were only "several hundred." The most optimistic view seems to be taken by Dr. John McCosker, director of the Steinhart Aquarium in San Francisco: "No one knows how many coelacanths there are, but there must be quite a few. I estimate that in modern times between 200 and 400 coelacanths have been caught off the Comoros."[158] Of course, no one, including McCosker, is likely to feel completely comfortable with "quite a few" as an estimate.

Without any direct information about population size we have to start with the incomplete catch data on the catch rate. Nearly everyone who has written about coelacanths over the years has referred to the catch rate by the Comoran fishermen as 3 to 5 per year. Indeed, this was probably true for the period from 1952, when the first Comoran specimen was recognized, to 1971. This is the interval for which the French group assembled catch statistics: 65 fish in twenty years, or 3.25 per year. However, most authorities also think that the number of coelacanths that has been caught has now risen to approximately 200. Thus, 135 fish have been caught in the last in the last eighteen years—about 7.5 per year. The catch rate has at least doubled. McCosker gives a higher outside total of 400 catches since 1952; if we were to allow 100 for the first twenty years and 300 for the second period, we would posit a tripling of the catch rate. Inevitably scientists fear the effects of such a growing pressure on the population.

Almost certainly the catch rate is largely a function of fishing effort: Increased effort gives a higher return, but by what proportion we do not know. If a doubling of the effort resulted in a full doubling of the catch, we could assume that the catch rate itself was not affecting population size. If the catch were to increase only by one-quarter when the effort is doubled, then we would know that the fishing pressure was decreasing the size of the stock (or at least that segment susceptible to fishing). For this reason, it would be revealing to know at what rate the fishermen had caught them and thrown them away before 1952. The best guess is that before 1952 specimens of *Latimeria* were being caught at the rate of one or two a year, that between 1952 and 1971 the rate increased to three to five a year, and that they are now being caught at six or seven a year, or more.

If the fishing effort has been reducing the number of individuals in the population, we would expect that the average age (and size) of specimens caught would slowly decline. In their survey of the data available up to 1971, Millot, Anthony, and Robineau concluded that the mean size of specimens being caught had not declined from 1938. Up to 1971 the smallest specimen taken had been eighty-five centimeters long. Millot, Anthony, and Robineau therefore concluded that the then-current primitive fishing methods, and a catch rate of three to five per year, were unlikely to deplete the stocks. But conditions have changed. The fishing pressure has increased, and more modern gear is being used, including fiber glass boats and outboard motors. Smaller specimens *are* now being taken along with bigger ones. Of the sixty-six catches recorded between 1938 and 1972, only one was less than a hundred centimeters long (1.5 percent). Of seventeen catches recorded between 1972 and 1977, five were less than a hundred centimeters (30 percent).[159] In 1973 the first obvious juvenile was taken, a fish only forty-three centimeters long. While the average size of the fish has changed little, the range has increased.

In view of the vulnerable pattern of viviparous reproduction, any change in the ratio of sexes in the catches should be viewed with concern. Up to 1972, of forty-four fish whose sex had been determined, twenty-five were male and nineteen female (and only very few of the females were in reproductive condition). Of fourteen fish reliably sexed between 1972 and 1977, seven were male and seven female. These numbers are small, but the shift may be significant. We desperately need catch data for the years since 1977.

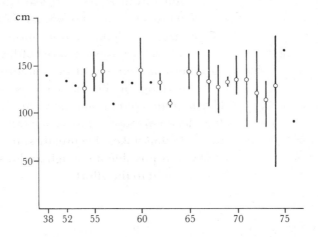

FIGURE 46   Size ranges of *Latimeria* caught between 1938 and 1976. Dots show single captures; circles show average for the year. Data from Millot, Anthony, and Robineau and from McCosker.

The biggest specimens we know about are all females, and because only one of the first seven specimens was a female (the sex of the very first specimen was never ascertained), Smith thought early on that the males and females had different distributions. He suggested that the females stayed deeper and only the males ventured close to the surface at night. This is not borne out by fuller statistics, but the possibility must be entertained that as they have created a coelacanth fishery over the years, the Comoran fishermen have come to sample a different portion of the population and that this accounts for the higher proportion of females and of smaller fish in the more recent catches. Once again this emphasizes our need to have more direct data on population size.

Finally, since the fish live to an considerable age, perhaps the full effect of the specimen capture from 1952 to 1990 has yet to be fully felt in the population statistics.

By all estimates, the pressure on the population(s) of *Latimeria* has increased over the years. There has been an active official market for the fish and a flourishing black market as well. As Comoran poverty increases, so will the pressure on the coelacanth. The main difficulty in assessing the situation is that no consistent account has been kept of the numbers of fish caught or of such data as size, sex, depth, and location since the French effort ceased. Therefore, it is a matter of highest priority to undertake underwater surveys of population size. Interestingly, the blotchy markings on the skin of the fish seem to be quite individualistic. One could potentially use this to identify individuals in underwater studies, just as one can identify whales from scars on the tail flukes. Eventually some sort of census of population size is possible if enough underwater observation time can be devoted to the effort.

CONSERVATION

The first person to have worried about the conservation status of *Latimeria* seems to have been none other than J. L. B. Smith.

He became worried about the situation when he realized that intense pressure for more catches would be generated as a result of the first observations of a live specimen (specimen number eight, see Chapter 4). He railed against the "senseless slaughter" by the Comoran fishermen and complained that the rewards being offered to the fishermen were only making the situation worse.[160] This is, of course, ironic since it was he who had instituted such bounties. But Smith was saying, enough is enough, we have achieved the main objective, and now the fishermen should no longer be encouraged to catch the fish, because the whole population might number only "a few hundred." That was 1956.

Since then the subject has been debated on and off. For example, Millot and Anthony proposed a moratorium at certain times of the year, in the light of the new evidence on reproduction. In the last couple of years the question of conservation has come to dominate debate about the coelacanth, in part because Fricke's successful use of a submersible to find and observe coelacanths has opened up a whole new range of technical possibilities for the capture of this fish.

Basically two opposing views clash on the conservation question. The conservative view, for which Fricke has been most outspoken, is that the fishing effort should be cut back and, preferably all fishing be halted, because of the danger to the species. Fricke and Dr. M. Bruton of the J. L. B. Smith Institute would like to see the fish protected within a Comoran Coelacanth National Park. The opposite view is again expressed by Dr. John McCosker, who is more bullish about population size and thinks that it is "completely off base to call coelacanths an endangered species until some realistic survey can be made. We haven't even sampled the waters off Mozambique, where they may also be found."[161] At present *Latimeria chalumnae* is not listed on the International Red List as an endangered species. Some progress, however, has been made recently. It was formerly listed under Article II of the Convention on International Trade in Endangered Species, which meant that trade in the fish was recognized to endanger its survival. But this was a weak statement, not enough to protect the fish, particularly against the strong illegal trade. Recently *Latimeria*'s status was revised to a listing under Appendix I, barring any trade. This is excellent progress and a credit to a new international organization, the Coelacanth Conservation Council/Conseil pour la Conservation du Coelacanth, a group formed expressly to conserve the fish and add it to the official list of the endangered.

Efforts to direct the activities of the local fishermen away from the coelacanth and bottom fishing toward a more pelagic (surface) fishery have also begun through an EEC fisheries project. There is very limited but nonetheless encouraging evidence that the catch rate may have started to fall in 1990.

I think everyone would agree on the need for a solid population count although no one has yet proposed to do it. But what should we do in the meantime—in the absence of good data? In view of all the uncertainty about the status of *Latimeria chalumnae* it seems to me that scientists must take the lead in following the most prudent course—namely, to *collect* as few new specimens as possible while concentrating on live *observations* (hoping that the use of submersibles itself does not disrupt the life of the stock and cause a population crash). While other populations might exist (probably not off Mozambique), *fishing on the Comores should stop,* including fishing by scientists and for scientific purposes.

Ironically, up to now the very fact of *Latimeria*'s interest to scientists has created the danger of its extinction. Scientific institutions, which in principle ought to be most sensitive to the problem, have continued to create the main market for the fish, usually in the form of frozen specimens. Japan's recent entry into the coelacanth research field has once again increased this pressure. As far as I can see, however, there is no scientific justification for further uncoordinated killing and collecting of specimens of *Latimeria.* Many questions remain about the biology of *Latimeria,* questions requiring properly prepared fresh tissues. For example, only such tissues will show whether the rostral organ has electroreceptive cells. In order to meet these needs, scientists should once again pool their requests, as we did in 1966 and 1972, and then meet

those requests with one or perhaps two fresh-caught specimens. There is actually no need at all for further formalin-
preserved specimens for exhibit. The world's museums already have a plentiful supply. There are at least twenty in the
Muséum Nationale d'Histoire Naturelle in Paris alone.

Furthermore, one can examine the published results resulting from the latest acquisitions of specimens by Japanese and
United States institutions and see that frankly, little new has
been learned except some molecular systematic data. For example, a lot is made of the use of CAT scanning of frozen
specimens to get anatomical data. But the embryo specimen
that was sent to San Francisco still has not been fully sectioned;
if that sectioning were completed, it would provide the most
minutely exacting anatomical details that one could wish for.

If, as I believe, one or two fresh specimens, dealt with coop-
eratively, could meet the demand for tissue samples for a long
time, work could concentrate on census and other nonintru-
sive observation by means of submersible vehicles or, even
more efficiently in certain cases, by remote cameras. There is a
lot of challenging work to be done and potentially exciting
results to be obtained. It would be extremely useful, for example, to have firsthand observations of feeding; presumably this
will soon be accomplished. Observations of mating behavior
and other aspects of reproduction would probably take longer
to make and would require some luck. Submersible exploration of the home range, in terms of depth and geographical
extent and population size, is perhaps the most pressing need.
This calls for both further exploration of the Comores and
exploration of similar habitats on neighboring island groups.
Indeed, to know the geographical range and population size is
essential if the potential extinction of this fish is to be avoided.

This need for restraint has to be set against the natural urge
of the Comorans to capitalize on a minor asset. It must be
emphasized that a great deal can be done to preserve the op-
tions by a voluntary restraint among those scientific organizations and individuals that would seek to own a coelacanth specimen. The situation may have been made even more difficult
by the practice of giving bounties to successful fishermen. In

the past such bounties have included not merely cash but, in the case of a Steinhart Aquarium trip to the Comores in 1975, the promise of a round-trip two-week visit to Mecca—"We had attracted a lot of fishermen this time," said its leader![162] Even if the bounties are stopped and the black market is controlled, without a ban on fishing, the Comoran fishermen will still catch *Latimeria* accidentally, as they always have through the years. This means that we need a ban on the *Ruvettus* fishery as well as a ban on fishing for *Latimeria* itself. At the very least the Comorans should not be encouraged to try to increase the take through bounties or donations of fiber glass boats and outboard motors aimed at the oilfish and coelacanth fishery. Considering the poverty of both the Comoran environment and the economy in general, this may be difficult, but one could always start by distributing the potential bounty money as a general subsidy.

However, other forces are at work in the marketplace for coelacanths. If the driving force of the fishing effort shifts (as it may already have done) toward institutions that principally seek to make money by public exhibition of a live coelacanth or toward individuals whose goal is to become famous as the "explorer" who first brought back a coelacanth alive, then I believe the risk to the population(s) will only grow, and grow rapidly. There is a possibility that a ban on trade and fishing involving *Latimeria* would make the Comoran government even more willing to grant permits for such "scientific purposes."

With some fascinating exceptions, a lot of the research that still needs to be done on *Latimeria* (painstaking plotting of ranges, for example) falls into the category of dull rote work, and as each new discovery is made, the glamour (and thus the potential for funding) attached to what is left to discover will inevitably decline. One does not have to be a cynic but merely a realist to see that this in turn will increase the pressure to accomplish something even more sensational.

Unfortunately there remains one major potential public relations coup to be pulled off: the capture of a live specimen for the purposes of keeping it alive for study and commercial display. Here opinions vary. On the technical side it is probably feasible. Originally the viability of live specimens played to the boat and dragged to shore seemed to indicate that *Latimeria* can survive rough handling. This view may have been exaggerated by Smith's embellishment of the story about the specimen from the Chalumna River "snapping viciously at various hands." But the 1954 and 1972 specimens survived for several hours in a semimoribund condition. Various groups, starting with the original French research team, have had plans that if a fish were brought to shore alive by a fisherman, it would be intercepted and transferred to a holding cage and resubmerged to the right depth. Here it would in theory recover from the effects of capture and any problems of depth or temperature change and then either be studied *in situ* by divers and cameras or be brought to the surface slowly and held under controlled conditions. Against this, however, is the growing information about the physiology of *Latimeria* suggesting that its energy reserves are drawn down rapidly. A spent fish that had been treated in this way might survive for many hours before succumbing but might not recover.

Apart from the trauma of capture, we can be sure that at least three factors may make it difficult, if not impossible, to keep *Latimeria* healthy at the surface for any length of time. As for any fish of this sort, they are pressure, temperature, and light.

The only realistic way to have a chance at keeping a live specimen alive for display and study would be to catch one *in situ* by luring it into a cage. The cage would then have to be tethered at the original depth until the next stage. One possibility would then be to bring it to the surface very slowly so that the fish could acclimatize calmly to depth changes, but in that case the temperature change might be lethal. We simply do not know. A more elaborate approach would be to transfer the specimen from the cage to a cooled and pressurized container while still at depth and then raise it to the surface. Only experimentation would tell whether a specimen of *Latimeria* would survive reduced surface pressures for any length of time. The fish would certainly have to be kept in the dark or in very low light because, the French observed, bright light bothers live *Latimeria* greatly.

The preceding is a bit like publishing directions for making a bomb. However, attempts to capture a live coelacanth are inevitable. Some of the technology is already available and improving all the time. Public interest is so great that there would be a plentiful financial return from a successful commercial display. There would probably be a sizable profit just from selling the films of the expedition trying to capture one, even if it were unsuccessful. However, it must be stated that the chances are great that many fish would be lost in the experimental phases of such a venture before one was successfully kept alive. If any one group were to succeed in exhibiting a live fish, there would be a rush of others who would want to try as well, and the pressures on the population would continue to escalate. The New York Aquarium and the Explorers Club of New York have already made joint attempts (in 1986 and 1987), which failed, but they were using the primitive approach of trying to catch a specimen by line. Since then Dr. Fricke has introduced the submersible as an observational tool, and (as he freely admits) obviously very little would be needed to make a submersible into a vehicle for capturing a specimen. Fricke has vehemently declared himself opposed to any such ventures, and until other groups try to use submersibles, the fish may be safe. However, in late 1989 a new set of players emerged in the form of a Japanese venture. The Japanese have developed a close, productive relationship with the Comoran government, with a joint program of study and more direct participation by the Comorans than any Western group has proposed. A "Japan Scientific Expedition of the Coelacanth" carried out two expeditions to the Comores in 1981 and 1983 and purchased three specimens from fishermen.[163] In late 1989 in what was reported to be a $1,760,000 project, a chartered ship headed for the Comores with the express intent of trying to capture a specimen alive for the Toba Aquarium.[164] Using underwater television surveillance, the expedition attempted to entice a coelacanth into a baited underwater cage that they planned to bring slowly to the surface. They were unsuccessful, but the escalation has begun.

What scientific purpose would be served by keeping a specimen in a tank for a few months? Obviously it would make someone a lot of money, but what could one learn that one could not learn from careful observation of live fish in their natural habitat? The answer is almost nothing, especially in comparison with the risks. For example, observing the fish feeding would be possible in both cases. A coelacanth in a tank could in theory be fed a herring or squid, like a sea lion at the zoo, but more could be learned in the field, in terms of prey selection, stalking behavior, use of the natural environment as a foraging ground, and so on. One could study the swimming mechanics (as I and others have studied shark swimming in large tanks), but more could be learned in the field where the coelacanths respond to currents, an uneven bottom, or other organisms. There is a great deal of scientific research that can be done only on a live fish, but one cannot imagine that serious manipulative experimentation would be allowed on a live specimen, so expensively obtained and maintained and so valuable as a source of income. How readily, for example, would the owners agree to have the fish anesthetized in order to draw blood or inject an experimental drug? What about implanting electrodes or catheters?

The Explorers Club and New York Aquarium's effort to catch specimens on lines and then keep them alive was justified as an effort to save fish that would be caught anyway:

We are attempting to override the Comoran government bounty on dead coelacanths by offering a higher premium for a live one.... Our procedure tests the ability of the coelacanth to recover from capture trauma if released by the fishermen into its natural habitat and thus paves the way for a national conservation policy in the Comores. If our handling techniques are successful, we will be able to transport breeding populations to safe areas, free from the intensive fishing off Grand Comore's southwest coast.[165]

In other words, these groups would encourage fishermen, through larger than usual bounties, to catch more fish and then try to keep them alive. Any fish that actually did survive would be transported to some other region where they are not currently known to live. All this sounds a bit like the notorious army unit in Vietnam that had to destroy the village in order to save it. It would be a lot simpler to offer the fishermen a smaller bounty not to fish at all, but who would pay money to see that? In fact, two groups of Explorers Club volunteers (paying four thousand dollars each) made trips to the Comores (1986, 1987), and they are reported to have purchased a total of six frozen specimens, which scarcely discourages the trade in fish.[166]

Incredibly, there is more scientific *folie de grandeur*. The argument has even been made that if *Latimeria* stocks really are very low, then capturing specimens and bringing them to an aquarium *to reproduce* may be the only way of ensuring the survival of the species. "Is it really wrong to try to study these creatures in captivity . . . and perhaps to breed them? The day may come when they need us as much as we need them," says Louis Garibaldi of the Explorers Club.[167] The club states: "Contrary to Dr. Fricke, we believe that survival of the coelacanth demands a program of human intervention as with the panda, the whooping crane, the California condor and other rare species."[168] In my opinion, this is a most improbable claim. The stock may well be threatened, but in light of what we currently know, to attempt to capture some or all of the survivors and keep them alive in a public aquarium will inevitably cause such losses before a single fish is exhibited, let alone pairs bred, that the exercise would significantly speed their extinction. It is simply premature (to put it mildly) to talk of breeding them in captivity when so far no one has yet kept one alive for more than a few hours. The technology, the knowledge, the information simply are not available, and such claims simply do not justify attempts at capture.

It seems extraordinary that *scientists* would be talking in these terms when we still have not managed to establish the most elementary piece of data: how many individuals there are. If there were tens of thousands of coelacanths, then commercial efforts to get live specimens for public display, with the inevitable losses, might be allowable, even encouraged. But let us turn the argument around. Suppose we knew for certain that only about a hundred coelacanths were left. Would it be better to "save" them using what crude and untested technology we have, or should we leave them alone, except for passive observation, until we had better confidence in our technology? If there really are only a hundred or even a thousand, should we have greater confidence in the coelacanth's own instincts and biology to ensure its reproduction, or should we capture them and put the survivors into tanks in New York or Tokyo? Apart from the ethical issues, as a scientist I have no doubt whatsoever that the currently available evidence requires that they be left alone. The greatest hope for their survival is to stop fishing for them because we do not yet have the technology to breed them in captivity. The magnitude of the problem is exemplified by the fact that the Japanese expedition planned to capture a male and a female for breeding. Now the sex of *Latimeria* specimens can be determined only by careful close examination—which means capture and handling. How many coelacanths would have to be caught before one had a male and a female? How many more would have to be caught before one had both a male and a female in breeding condition? How many fish in breeding condition would have to be caught before breeding was successful? How many successful breedings would be necessary before young survived to maturity? And so on.

If *Latimeria* is to be safely brought up to the surface and kept alive, let alone bred, then we need to experiment so that we can first do it with *Ruvettus* or *Hexanchus* and understand enough of the coelacanth's physiology to counter the effects of pressure, temperature, and light. None of that is difficult; perhaps being patient is what is difficult. Unfortunately in this age of instant gratification it may be hard to find people to tackle the unglamorous preliminary work, and even harder to get funding.

Finally, we should ask the ethical question, How important is the survival of the species *Latimeria chalumnae*? It is only a fish after all, and inedible at that. This is a question one can ask about any species, including our own. Fifty years ago we did not know that a population of living coelacanths existed, so who would care if it were gone after fifty more years? But we do know this species, of course, and it is a bit like the problem of a stray kitten. Having taken it into the house, we have assumed responsibility for it.

There are several obvious reasons why we should ensure the survival of the species. We found this fish by chance, and we have a moral obligation to pass it on for future generations to see and know. We have a moral obligation to face our own ignorance and to cease practices that may be endangering the survival of this fish. If the species needs human intervention, to survive, as it patently does not, we would have to be sure that we know what we are doing. *Latimeria chalumnae* does not present an obstacle to any human endeavor, so we are not in the position of trading its extinction for some tangible human benefit. It is not in our way; it existence does not prevent the building of a dam or the feeding of starving children. We do not need its habitat to build naval bases or to test pesticides. Nor, as far as we can tell, does *Latimeria* offer any tangible economic opportunity (except for a few would-be showmen). Its oil has no apparent medicinal value (although that of *Ruvettus* does, but no one is interested in *Ruvettus*). The oil really is not an aphrodisiac, nor does it cause liver cancer; all that is a stupidity. In a very practical sense, we need to make sure that *Latimeria* does not become extinct simply because our human species likes to know things and there is so much that we still do not know about it.

We have a moral obligation not to be careless with something that belongs to someone else. If we were dealing with real property—a car, an oil well—we would have a legal obligation as well. But the living coelacanth does belong to someone else. Principally it belongs to future generations, generations of Comoro islanders as well as Americans, British, or Japanese, to the public as well as to scientists. We few scientists (there are probably fewer than fifty of us) who actively study coelacanths and the additional twenty or so entrepreneurs who want to exploit this fish (in a real sense scientists exploit it, too, of course) are scarcely majority stockholders.

Of all the endangered and threatened species in the world, *Latimeria chalumnae* may well be the only organism whose extinction is by scientists. There are quite a few species of orchids and shells that have been brought to (or close to) extinction by collectors after scientists have made them visible, but in the case of *Latimeria* essentially the whole catch has ended up in the control of scientists. The species was discovered by scientists and collected for scientists, for research and (so far at least) for exhibit in scientific institutions. This is another historic first for science, although one of dubious distinction. It puts a special onus on scientists to ensure its survival: If *Latimeria* becomes extinct, there really will be no one else to blame.

The easiest way to accommodate to the extinction of a species is if it happens "by accident." But as we know, most accidents are caused, usually by greed or stupidity. Once the possibility of simply leaving the fish alone has been articulated, it is hard to avoid. There is nothing to be lost by taking one or two last culls for carefully designed research and then totally banning fishing. The world's public might have to wait for a few years before having the thrill of seeing a live fish confined in a tank. In the meantime they would have to put up with fabulous television pictures taken of fish behaving normally in their natural environment.

Latimeria chalumnae is really just another a fish. But . . .

Notes

1. M. Courtenay-Latimer. 1979. My story of the first coelacanth. Occ. Pap. Calif. Acad. Sci. 134:6–10.

2. M. Courtenay-Latimer. 1989. In: Remembering the coelacanth: a 50th anniversary perspective, ed. R. Greenwell. Internat. Soc. Cryptozool. Tucson.

3. M. Courtenay-Latimer, 1979. op. cit., p. 7.

4. J. L. B. Smith. 1956. *Old Fourlegs: The Story of the Coelacanth.* London: Longman, Green.

5. J. L. B. Smith. 1940. A living coelacanth fish from South Africa. Trans. Roy. Soc. S. Af. 28:1–106.

6. H. Goosen. 1989. In: Remembering the coelacanth: a 50th anniversary perspective, loc. cit.

7. H. Goosen, 1989, op. cit., p. 16.

8. M. Courtenay-Latimer, 1989, op. cit., p. 16.

9. C. Munnion. 1988. Remembering old fourlegs. Optima 36:42–51.

10. J. L. B. Smith, 1956, op. cit., p. 27.

11. A. Smith Woodward. 1898. *Catalogue of Fossil Fishes of the British Museum (Natural History).* Volume II. London: BMNH.

12. J. L. B. Smith, 1956, op. cit., p. 31–32.

13. Ibid., p. 32.

14. Ibid., p. 35.

15. Ibid., p. 31.

16. Ibid., p. 37.

17. M. Courtenay-Latimer, 1979, op. cit., p. 9.

18. J. L. B. Smith, 1956, op. cit., p. 41.

19. Margaret Smith. 1970. The search for the world's oldest fish. Oceans 3 (6):26–36.

20. J. L. B. Smith, 1939. A living fish of Mesozoic type. Nature 143:455–456.

21. E. I. White. 1939. One of the most amazing events in the realm of natural history in the twentieth century. London Ill. News. March 11, 1939, Supplement.

22. J. R. Norman. 1939. A living coelacanth from South Africa; a fish believed to have been long extinct. Proc. Linnean Soc. London. 151:142–145.

23. Survivor of an ancient line. The Times, London, March 10, 1939. A coelacanth fish. The Times, London, March 17, 1939.

24. J. L. B. Smith, 1956, op. cit., p. 51.

25. Ibid., p. 53.

26. E. I. White, 1939. op. cit., p. 2.

27. J. L. B. Smith. 1939. op. cit., p. 456.

28. J. R. Norman, 1939, op. cit., p. 143.

29. J. L. B. Smith, 1956, op. cit., p. 54.

30. Ibid., p. 55.

31. J. L. B. Smith, 1940, op. cit., p. 53.

32. A. Smith Woodward, 1940. The surviving crossopterygian fish, *Latimeria.* Nature 239:283–285.

33. M. Courtenay-Latimer, 1989, op. cit., p. 13.

34. J. L. B. Smith, 1939, op. cit., p. 1.

35. J. L. B. Smith, 1956, op. cit., p. 75.

36. J. L. B. Smith, 1949. *Sea Fishes of Southern Africa.* South Africa: Central News Agency.

37. J. L. B. Smith, 1956, op. cit., p. 84.

38. Ibid., p. 85.

39. Ibid., p. 89.

40. Ibid., p. 101.

41. Ibid., opp. cit. p. 109.

42. Prehistoric fish believed caught. New York Times, December 28, 1952. Air race to save dead fish stirs scientists here. New York Times, December 30, 1952. 14-year hunt yields "missing link" fish. New York Times, December 30, 1952. Scale of "Missing link" fish given to Malan, foe of evolution theory. New York Times, December 31, 1952. No title. Le Monde, January 1, 1953. Scientist tells of rare fish find. New York Times, January 2, 1953.

43. Shirley Bell. 1969. *Old Man Coelacanth.* Johannesburg: Vortrekkerpers, p. 111.

44. Affane Mohamed, 1965. Capture du 2ème coelacanth "identifié" le 22 decembre 1952 à Domoni (Anjouan). Unpublished affidavit, privately circulated by Professor J. Millot.

45. J. L. B. Smith, 1956, op. cit., p. 146.

46. Ibid., p. 153.

47. J. L. B. Smith, 1953. The second coelacanth. Nature 171:99–101.

48. J. Millot, 1955. Unité specifique des coelacanthes actuels. La Nature 3238: 58–59.

49. J. Dugan, 1955. The fish. Collier's, September 16. 64–68.

50. J. L. B. Smith, 1956, op. cit., p. 211.

51. Ibid.

52. J. Anthony, 1976. *Operation Coelacanthe.* Paris: Arthaud.

53. J. Millot. 1953. Notre coelacanthe. Revue Madagascar 17:18–20. J. Millot. 1954. Le Troisième Coelacanthe. Historique éléments d'Écologie, morphologie externe, documents divers. Naturaliste Malagache, Suppl. 1.

54. J. L. B. Smith, 1956, op. cit., p. 216.

55. J. Anthony, 1976, op. cit.

56. J. Dugan, 1955, op. cit., p. 66.

57. Ibid., p. 67.

58. Ibid.

59. J. Millot. 1955. First observations on a living coelacanth. Nature 175:362–363.

60. J. Millot, J. Anthony, and D. Robineau. 1972. État commente des captures de *Latimeria chalumnae* Smith (Poisson, Crossopterygien, Coelacanthide) effectués jusqu'au mois d'Octobre 1971. Bull. Mus. Hist. Nat. Paris. 53:533–548.

61. J. Atz. 1976. *Latimeria* babies are borne, not hatched. Underw. Nat. 9:4–7.

62. C. R. Darwin, 1859. *On the Origin of Species*. . . . London: Murray.

63. R. R. Hessler, 1984. In: *Living Fossils.* ed. N. Eldredge and S. M. Stanley. New York: Springer-Verlag.

64. E. B. Conant. 1986. An historical discussion of the literature of Dipnoi: Introduction to the bibliography of lungfishes. In *The Biology of Lungfishes,* W. E. Bemis, W. W. Burggren, and N. E. Kemp. New York: A. R. Liss, ed.

65. L. Agassiz. 1836. *Recherches sur les Poissons Fossiles.* Neuchâtel.

66. E. A. Stensio, 1937. On the Devonian coelacanthids of Germany with especial reference to the dermal skeleton. K. Svenska VetenskapAkad. Handl. ser. 3, 16:1–56.

67. R. Lund and W. L. Lund. 1975. Coelacanths from the Bear Gulch Limestone (Namurian) of Montana and the evolution of the Coelacanthiformes. Bull. Carnegie Mus. Nat. Hist. 25:1–74.

68. D. Raup. 1986. *The Nemesis Affair.* New York: Norton.

69. J. A. Moy-Thomas and R. S. Miles. *Palaeozoic Fishes.* New York: Saunders.

70. J-P. Lehman. 1952. Étude complémentaire des poissons de l'Eotrias de Madagascar. K. Svenska VetenskapAkad. Handl. 4 (2) 1–201.

71. J. G. Maisey. 1986. Coelacanths from the Lower Cretaceous of Brazil. Novitates Am. Mus. Nat. Hist. 2866:1–26.

72. B. Schaeffer. 1952. The Triassic coelacanth fish *Diplurus* with

observations on the evolution of the Coelacanthini. Bull. Am. Mus. Nat. Hist. 135:287–432.

73. D. M. S. Watson, 1927. The reproduction of the coelacanth fish, *Undina.* Proc. Zool. Soc. London 19827:453–457.

74. E. B. Conant, op. cit.

75. E. D. Cope, 1892. On the phylogeny of the Vertebrata. Proc. Am. Phil. Soc. 30:278–281.

76. J. T. Wilson, 1972. *Continents Adrift.* Readings from American Scientist. San Francisco.

77. P. Molnar and J. Franchetau. 1973. Relative motion of hotspots in the mantle. Nature 246:288.

78. J. D. Dana, 1894. *Manual of Geology.* New York: American Book Company.

79. C. M. Emerick and R. A. Duncan. 1982. Age progressive volcanism in the Comores Archipelago, western Indian Ocean and implications for Somali plate tectonics. Earth and Planetary Sci. Lett. 60:415–428.

80. M. Griffin. 1986. The perfumed isles. Geog. Mag. 58524–527. J. F. G. Lionnet. 1983. Islands not unto themselves. Ambio 12:288–289. J. M. Ostheimer. 1973. Political development in the Comores. Afr. Rev. 3:491–506.

81. P. Scoones. 1980. Coelacanth encounter. Skin Diver 29:8–9.

82. J. Millot and J. Anthony. 1960–1978. *L'Anatomie de Latimeria chalumnae.* 3 vols. Centre Nat. Res. Sci. Paris.

83. J. E. McCosker. 1979. Inferred natural history of the living coelacanth. Occ. Pap. Calif. Acad. Sci. 134:17–24. E. K. Balon, M. N. Bruton, and H. Fricke. 1989. A fiftieth anniversary reflection on the living coelacanth, *Latimeria chalumnae:* some new interpretations of the natural history. Env. Biol. Fishes 23:241–280.

84. J. Millot. 1958. *Latimeria chalumnae,* dernier des crossopterygiens. In: *Traité de Zoologie,* ed. P. Grasse. Vol. 13. Paris: Masson.

85. S. M. Andrews. 1977. The axial skeleton of the coelacanth. In: *Problems in Vertebrate Evolution,* ed. S. M. Andrews, R. S. Miles, and A. D. Walker. London: Academic Press.

86. J. Millot, 1955. op. cit.

87. J. Millot and J. Anthony. 1960. Appareil génital et reproduction des coelacanthes. C. R. Hebd. Séanc. Acad. Sci. Paris D 251:442–443.

88. J. Millot and J. Anthony. 1960. Le plus vieux poisson du monde. Sciences 6:7–20.

89. J. Millot, J. Anthony, and D. Robineau, 1972, op. cit.

90. K. S. Thomson. 1967. Mechanisms of intracranial kinetics in fossil rhipidistian fishes (Crossopterygii) and their relatives. J. Linn. Soc. London Zool. 46:223–253.

91. K. S. Thomson. 1986. A fishy story. Amer. Sci. 74:169–171.

92. J. Anthony, 1976, op. cit., p. 89.

93. G. R. Forster, J. R. Badcock, N. R. Merret, M. R. Longbottom, and K. S. Thomson. 1974. Results of the Royal Society Indian Ocean deep slope fishing expedition. Proc. Roy. Soc. London B 175:367–404.

94. K. S. Thomson. 1981. The capture and study of two coelacanths off the Comoro Islands, 1972. Nat. Geog. Soc. Res. Rep. 13:615–622.

95. J. Anthony and J. Millot. 1972. Première capture d'une femelle de coelacanthe en état maturité sexuelle. Séances acad. Sci. Paris D 274: 1925. J. Millot and J. Anthony. 1974. Les Oeufs du coelacanthe. Sci. Nat. Paris 121:3–4.

96. N. A. Locket and R. W. Griffith. 1972. Observations on a living coelacanth. Nature 237:175.

97. H. Fricke. 1988. Coelacanths. The fish that time forgot. National Geog. 173:824–838. H. Fricke, O. Reinicke, H. Hofer, and W. Nachtigall. 1987. Locomotion of the coelacanth *Latimeria chalumnae* in its natural habitat. Nature 329:331–333.

98. J. C. Nevenzel, W. Rodegker, J. F. Mead, and M. S. Gordon. 1966. Lipids of the living coelacanth, *Latimeria chalumnae*. Science 152: 1753–1755.

99. N. A. Locket, 1980. Some advances in coelacanth biology. Proc. Roy. Soc. London B 208:265–307.

100. H. Fricke and R. Plante. 1988. Habitat requirements of the living coelacanth *Latimeria chalumnae* at Grande Comore, Indian Ocean. Naturwiss. 75:149–151.

101. H. Fricke et al., 1987, op. cit.

102. M. Bruton. 1989. The coelacanth—can we save it from extinction? World Wildlife Fund Reports, October/November 1989. 10–12. H. Fricke. Quoted New York Times, March 22, 1988.

103. M. Menache. 1954. Étude hydrogéologique sommaire de la région d'Anjouan, en rapport avec le pêche des coelacanthes. Mem. Inst. Sci. Mad. ser. A,9:152–185.

104. G. R. Forster. 1974. The ecology of *Latimeria chalumnae*. J. L. B. Smith: Results of field studies from Grande Comore. Proc. Roy. Soc. London B 186:291–296.

105. H. Fricke and R. Plante, 1988, op. cit., p. 150.

106. G. R. Forster, 1974, op. cit.

107. H. Fricke, and R. Plante, 1988, op. cit., p. 150.

108. D. De Sylva. 1966. Mystery of the silver coelacanth. Sea Frontiers 12:172–175.

109. J. Anthony, 1976, op. cit., p. 165.

110. H. Fricke, quoted in New York Times, March 22, 1988.

111. K. S. Thomson. 1966. Mobility of the skull and fins in the coelacanth, *Latimeria chalumnae*. Am. Zool. 6:565–566.

112. J. Millot, 1955, op. cit.

113. N. A. Locket, and R. W. Griffith. 1972. Observations on a living coelacanth. Nature 237:175.

114. H. Fricke et al., 1987, op. cit.

115. K. S. Thomson. 1969. The biology of the lobe-finned fishes. Biol. Rev. 44:91–154.

116. B. Dean. 1906., Notes on the living specimens of the Australian lungfish, *Ceratodus forsteri*, in the Zoological Society's collection. Proc. Zool. Soc. London 1906:168–1787.

117. H. Fricke et al., 1987, op. cit.

118. K. S. Thomson. 1966. Intracranial mobility in the coelacanth. Science 153:999–1000.

119. K. S. Thomson, 1973. New observations on the coelacanth fish, *Latimeria chalumnae*. Copeia 1973:813–814.

120. K. S. Thomson, 1966. op. cit.

121. D. Robineau and J. Anthony. 1973. Bioméchanique du crâne de *Latimeria chalumnae* (Poisson, Crossopterygien, Coelacanthide) C. R. Hebd. Séanc. Acad. Sci. Paris D 276:1305–1308.

122.   G. Lauder. 1980. The role of the hyoid apparatus in the feed-
       ing mechanism of the coelacanth *Latimeria chalumnae.* Copeia
       1:1–9.

123.   R. Nieuwenhuys, J. P. M. Kremers, and C. van Huijzen. 1977.
       The brain of the crossopterygian fish *Latimeria chalumnae.* Anat.
       Embryol. 151:157–169.

124.   J. Millot and J. Anthony. 1956. L'organe rostral de *Latimeria*
       (Crossopterygien, Coelacanthide). Annls. Sci. Nat. B 28:381–
       388.

125.   K. S. Thomson, 1977. On the individual history of cosmine
       and a possible electroreceptive function for the pore-canal sys-
       tem in fossil fishes. In: *Problems in Vertebrate Evolution,* loc. cit.

126.   A. J. Kalmijn. 1978. Electric and magnetic sensory world of
       sharks, skates and rays. In: *Sensory Biology of Sharks, Skates, and
       Rays,* ed. E. S. Hodgson and R. F. Matthewson. Washington,
       D.C.: Govt. Printing Office.

127.   W. E. Bemis and T. E. Hetherington. 1982. The rostral organ
       of *Latimeria chalumnae*: Morphological evidence of an elec-
       troreceptive function. Copeia 1982:467–471.

128.   H. Fricke and R. Plante, 1988, op. cit.

129.   G. E. Pickford and F. B. Grant. 1967. Serum osmolarity in the
       coelacanth *Latimeria chalumnae*: urea retention and ion regula-
       tion. Science 155:568–570. R. W. Griffith. 1980. Chemistry of
       the body fluids of the coelacanth, *Latimeria chalumnae.* Proc.
       Roy. Soc. London B 208:329–347.

130.   J. L. B. Smith. 1953. Problems of the coelacanth. S. Afr. J. Sci
       49:279–281.

131.   G. W. Brown and P. P. Cohen. 1960. Comparative biochemis-
       try of urea synthesis. 3. Activities of urea-cycle enzymes in vari-
       ous higher and lower vertebrates. Biochem. J. 75:82–91.

132.   G. W. Brown and S. W. Brown. 1967. Urea and its formation in
       coelacanth liver. Science 155:570–572.

133.   R. W. Griffith, 1985. Habitat, phylogeny, and the evolution of
       osmoregulatory strategies in primitive fishes. In: *Evolutionary
       Biology of Primitive Fishes,* ed. R. E. Foreman, A. Gorbman, J. M.
       Dodd, and R. Olsson. New York: Plenum.

134.   S. Ohno. 1970. *Evolution by Gene Duplication.* New York:
       Spinger-Verlag.

135. R. A. Pedersen. 1971. DNA content, ribosomal gene multiplicity, and cell size in fish. J. Exp. Zool. 177:65–78.

136. K. S. Thomson. 1972. An attempt to reconstruct evolutionary changes in the cellular DNA content of lungfish. J. Exp. Zool. 180:363–372. K. S. Thomson and K. Muraszko. 1978. Estimation of cell size and DNA content in fossil fishes and amphibians. J. Exp. Zool. 205:315–320.

137. M. Vialli. 1957. La quantita di acido desossiribonucleico per nucleo negli eritociti di Latimeria. Ist. Lombardo (Rend. Sc.) 91:680–685.

138. K. S. Thomson, J. G. Gall, and L. W. Coggins. 1973. Nuclear DNA contents of coelacanth erythrocytes. Nature 241:126.

139. M. C. Cimino and G. F. Bahr. 1973. Nuclear DNA content and chromatin ultrastructure of the coelacanth Latimeria. J. Cell. Biol. 59:55.

140. G. M. Hughes. 1979. Ultrastructure and morphometry of the giulls of Latimeria chalumnae and a comparison with the gills of associated fishes. Proc. Roy. Soc. London B.

141. G. M. Hughes and Y. Itazawa. 1972. The effect of temperature on the respiratory function of coelacanth blood. Experientia 28:1247.

142. J-C. Hureau and C. Ozouf. 1977. Détermination de l'age et croissance du coelacanthe Latimeria chalumnae Smith, 1939 (Poisson, Crossopterygien, Coelacanthide). Cymbium 2:129–137.

143. W. S. Hoar. 1969. Reproduction. In: Fish Physiology, ed. W. S. Hoar and D. J. Rondell. Vol. 3. New York: Academic Press.

144. P. H. Greenwood, The natural history of African lungfishes, and A. Kemp. The biology of the Australian lungfish. 1986. In: Biology of Lungfishes, loc. cit.

145. J. Millot and J. Anthony, 1960, op. cit.

146. J. Millot and J. Anthony. 1974. Les oeufs du coelacanthe. Science et Nature 121:3–4.

147. H-P. Schultze. 1972. Early growth stages in coelacanth fishes. Nature 236:90–91. H-P. Schultze. 1980. Eier legende und lebend gebärende Quastenflosser. Natur und Museum 110:93–124.

148. R. W. Griffith and K. S. Thomson. 1973. *Latimeria chalumnae*: Reproduction and conservation. Nature 242:617–618.

149. C. L. Smith, C. S. Rand, B. Schaeffer, and J. Atz. 1975. *Latimeria*, the living coelacanth, is ovoviviparous. Science 190:-1105–1106. *Envir. Biol. of Fishes.* In press.

150. N. A. Locket. 1972. A future for the coelacanth? New Scientist 570:546–458.

151. J. P. Wourms, J. W. Atz, and M. D. Stribling. 1990. Viviparity and the maternal-embryonic relationship in the coelacanth *Latimeria chalumnae. Envir. Biol. of Fishes.* In press.

152. S. M. Andrews and T. S. Westoll. 1970. The post-cranial skeleton of *Eusthenopteron foordi* Whiteaves. Trans. Roy. Soc. Edinb. 68:207–329.

153. M. D. Lagios. 1979. The coelacanth and the Chondrichthyes as sister groups: A review of shared apomorph characters and a cladistic analysis and reinterpretation. Occ. Pap. Calif. Acad. Sci. 134:25–44.

154. K. S. Thomson. 1980. The ecology of Devonian lobe-finned fishes. In: *The Terrestrial Environment and the Origin of Land Vertebrates,* ed. A. L. Panchen London: Academic Press.

155. K. S. Thomson, 1969, op. cit.

156. G. C. Packard. 1974. The evolution of air-breathing in Paleozoic gnathostome fishes. Evolution 28:320–325.

157. B. Dean, 1906, op. cit., figure 55.

158. J. E. McCosker, quoted in M. W. Browne. 1988. Conserving fossil fish. Aquariums, June 1988.

159. J. E. McCosker, 1979, op. cit.

160. Senseless slaughter of rare fish assailed. New York Post, June 5, 1956.

161. J. E. McCosker, in Browne, 1988, op. cit.

162. J. L. Hopson. Fins to feet to fanclubs: An (old) fish story. Science News 109:28–30.

163. N. Suzuki, Y. Suyehiro, and T. Hamada. 1985. Initial reports of expeditions for coelacanths—Part I Field Studies in 1981 and 1983. Sci. Pap. Coll. Arts Sci. Univ. Tokyo 35:37–79.

164. Effort to capture "fossil fish" draws fire. New York Times, September 12, 1989.

165. J. Hamlin. 1989. Letter to New York Times, March 24, 1989. In
      response to article Do scientists pose a threat to rare "fossil
      fish"? New York Times, March 22, 1989.

166. M. Hall. 1989. The survivor. Harvard Magazine. 91 (1):36–42.

167. L. E. Garibaldi, quoted in Browne, 1988, op. cit.

168. J. Hamlin, 1989, op. cit.

Index

Acanthostega, 206
Actinistia, 78
    see also Coelacanthini
Actinopterygii, 81, 83
Agassiz, Louis, 77
air bladder, *see* lungs
Air Comores, 68
Aldabra, 104, 120
American Museum of Natural
    History, 53, 68, 198
Amia, 74, 82
Amirantes, 4, 48
Amphibia, 43, 206, 211
Amphiuma, 187
ampullae of Lorenzini, 175
anglerfish, 167
Anjouan, 53, 59, 65, 101, 104, 106,
    121
    see also Comore Archipelago
Anthony, Jean, 65, 121, 151, 221,
    224
Archaeopteryx, 90
Aristea, 22
Assumption, 104
asteroid theory of extinction, 85

Atz, James, 201
Australia, 218
Axelrodichthys, 87, 89, 97

Barnard, K. H., 27, 30, 32–33,
    68
Bassas da India, 149
Bilbao, 149
Bird Island, 20–21
birds, 43, 183
Black Sea, 95
blood chemistry, 177
    see also Latimeria, physiology
blue shark, *see Prionace*
bony fishes, *see* Osteichthyes
Bourou, Ahmed Hussein, 52, 55–56,
    60
bowfin, *see Amia*
British Broadcasting Corporation,
    109
British Museum (Natural History),
    37, 41
Brown, George and Susan, 181
Bruce-Bays, J., 25, 41
Bruton, Michael, 145, 224

cancer, liver,
cannibalism, in utero, 201
Cape Town, 49
Carcharodon albimarginatus, 121
Castor Bank, 104, 120
castor-oil fish, *see Ruvettus*
catfish, walking, *see Clarias*
Center, Robert, 25, 40
Centroscymnus, 141
Cephalocarida, 73
Chagos, 102
Chagos-Laccadive Ridge, 148
Chalumna River, 19, 21, 23–24
*chalumnae, see Latimeria; Latimeria
chalumnae*
character analysis, 205–40
Children's Hospital, San Francisco,
199
choana, *see* nasal organ
Clarias, 163
coelacanth, living, *see Latimeria*
coelacanth, silver icon, 149–52
Coelacanthini, naming, 74, 78
characters of, 88
fossil record, 68–94
coelacanths, general, *see*
Coelacanthini
Coelacanthus granulatus Agassiz,
77
Comoran government, 117
Comore Archipelago, 48, 51, 59, 61,
99, 104, 146
Comore hot spot, 100
Comores (Comoro Islands), *see*
Comore Archipelago
conservation, *see Latimeria,*
conservation
continental drift, 95–97, 148
Convention on International Trade
in Endangered Species, 221
Cordelière Bank, 104, 120
Cosmoledo, 4, 104
Coteur, Dr. le, 53
Coudert, Pierre, 54, 57
Council for Scientific and Industrial
Research, 45, 59, 63
Courtenay-Latimer, Marjorie,
20–44, 68, 74, 123, 140, 177

Copenhagen, 68
Crossopterygii, 29, 41–43, 91–92
cytochrome c, 187

Darwin, Charles, 71, 121
dawn redwood, *see Metasequoia*
deep sea, 39, 43, 138
see also *Latimeria,* distribution and
environment
Deep Slope Fishing Expedition,
120
desoxyribonucleic acid (DNA),
183–86
Didelphis, 73
Diégo-Suarez, 60, 100
Diplocercides, 78, 94
Diplurus, 85, 87
Dipnoi, 25, 33, 72, 91–92, 143, 163,
181, 184, 212, 216
Doddington, 20
Domoni, 52–53, 56
see also Comore Archipelago
Drury, Mr., 41
Dunnotar Castle, 49–51
Durban, 50–53
Dutch Reformed Church, 61
Dzaoudzi, 50, 57, 60, 63, 1–4

East London, 20, 30, 41
East London Dispatch, 36–37
eggs, *see* reproduction, general
see also *Latimeria,* reproduction
electroreception, 174–76
see also *Latimeria,* brain and senses
endangered species, *see Latimeria,*
conservation
Etelis, 120
Eusthenopteron, 206
evolution, 61
see also *Latimeria,* relationships;
tetrapods, origins
Explorers Club, 230–31
extinction, 85
eye, *see Latimeria,* brain and senses

Farquahar, 100, 148
fertilization, 194–98
see also *Latimeria,* reproduction

fins, 24, 63, 88, 158, 207
   see also Latimeria, fins and
      swimming
fishing, Comores, 106–08, 120, 136,
   226–27
Forster, G. R., 120, 147
Fourmanoir, Pierre, 64
freshwater, undersea, 146
Fricke, Hans, 124–25, 130, 140–45,
   150, 160–61, 168, 176, 182,
   224, 231
frog, see Rana

Garibaldi, Louis, 231
Garrouste, Georges, 65–66, 69,
   198
gars, see Lepisosteus
geological time scale, 74
Geyser Bank, 67, 100, 104
gill mechanics, 164, 169–70, 180
   see also Latimeria, respiration and
      metabolism
Gnathostomata, 81, 164
gombessa (ngombessa) 55
Goosen, Hendrik ("Harry"), 22–24,
   29, 42, 66
Grahamstown, 34, 50
Grande Comore, 65, 67, 100, 104,
   106, 121
   see also Comore Archipelago
Greenland, 218
Griffith, Robert ("Bob"), 123, 183,
   196
growth rings, 188
   see also Latimeria, age

hagfish, 79, 178–79
Hawaii, 99–100
"head stand," see Latimeria, behavior
hemoglobin, 187
   see also Latimeria, respiration and
      metabolism
Hexanchus, 147, 199, 232
Hiariako, 67
horseshoe crab, see Limulus
hot spots, 99
Hughes, George, 186
Hunt, Eric, 48–54

Hussein, Ahmed, see Bourou,
   Ahmed Hussein
Hussein, Houmadi, 65
Huxley, T. H., 145

Iceland, 95, 99
Ichthyostega, 206, 211–14
Iconi, 123
   see also Comore Archipelago
Illustrated London News, 37, 39, 62
Indian Ocean, 19, 139, 147–48
   history, 94–102
intracranial joint, 42, 88, 110, 116,
   166, 216
   see also Latimeria, feeding

jaws, origin of, 164
Japan, Scientific Expedition of the
   Coelacanth, 226–29

Kaar, Madi Yousouf, 123
Kalmijn, A. J., 175
Karthala, Mount, 104, 146
Kenya, 47
Keynes, Quentin, 67
Kirsten, Mr., 25
Knysna, 28

Laccadives, 44, 148
lamprey, 79
Latimeria, age, 187–89
   behavior, 110, 113
   brain and senses, 172–76
   catch data, 104–14, 137–41,
      230–23
   color, 23–24, 26, 198
   conservation, 223–34
   depth, 137–41
   distribution and environment,
      104–14, 135, 142–52
   DNA, 185
   eye and vision, 141–41
   feeding, 110, 164–70
   fins and swimming, 24, 113, 124,
      156–64
   frozen specimen, Yale, 118–19
   intracranial joint, 42, 88, 100, 110,
      116, 166, 216

Latimeria (continued)
live specimens, 108–09, 123, 125,
159–60, 227–33
lung, 39, 89, 132, 212
notochord and vertebral column,
110
photographs, 35, 108–09
physiology, 176–87
population size, 219–23
relationships, 202–12
reproduction, 114, 121–22,
190–201
respiration and metabolism,
186–87
scales, 24, 26, 33, 139, 188–89
size, 114, 222
stomach contents, 100
temperature, 142
Yale specimen, 123, 168, 172,
176–77, 182
Latimeria chalumnae, named, 36, 39
Lauder, George, 170
Laugia, 88
Laurenço Marques, 57
Lehr, André, 66
Lepidosiren, 75, 92
Lepisosteus, 75, 82
Leven Bank, 104, 120
Life magazine, 108
limbs, tetrapod, 207–08, 214–15
Limulus, 71
Linnaeus (Karl von Linné), 77
Linnean Society of London, 37–38,
40
live-bearing, *see* ovovivarity;
viviparity
see also Latimeria, reproduction
living fossils, 71–78, 143
lobe-finned fishes, *see* Sarcopterygii
Loch Ness Monster, 151
Locket, Adam, 123
longevity, fishes, 189
long lines, 40, 138
see also fishing, Comores
Lumbo, 57
lungfish, African, *see* Protopterus
lungfish, Australian, *see* Neoceratodus
lungfish, general, *see* Dipnoi

lungfish, South American, *see*
Lepidosiren
lungs, 39, 89, 132, 139, 212
see also Latimeria, lungs

McCosker, John, 220–24
Macropoma, 31, 37, 78, 85, 87, 89
Madagascar, 44, 46, 48, 53, 60, 87,
99, 100, 145, 148
Malan, Daniel, 51–52, 59, 61, 67
Malania, 58, 62–63
Maldives, 44
mammals, 43, 183, 194
marsupials, 194
Mascarene Ridge, 44, 101, 148
Mauritius, 44, 148
Mayotte, 55, 100, 104–05
see also Comore Archipelago
Mediterranean Sea, 95
Metasequoia, 71
Mid-Atlantic Ridge, 96–97
midwife toad, *see* Xenopus
Millot, Jacques, 54, 64–66, 221, 224
missing link, 71
Mohamed, Affane, 53–59
Mohamed, Zema ben Said, 66
Mohéli, 101, 104
see also Comore Archipelago
molecular evidence, 209
Mombasa, 49
monotremes, 194
Moroni, 67
see also Comore Archipelago
Mozambique, 46, 57, 64, 104
Mozambique Channel, 44–45, 144,
149
mud puppy, *see* Necturus
mudskipper, *see* Periophthalmus
Muséum National d'Histoire
Naturelle, 65, 108–109, 120–21,
226
Mutsamudu, 53, 55–56, 64
see also Comore Archipelago
Myctophidae, 110–11

Nairobi, meeting, 64
nasal organ, 39, 42
see also Latimeria, brain and senses

National Academy of Sciences, 108,
    120–21
National Geographic Society, 121
Nature, 36, 37, 39, 42, 62
N'duwaro, 48, 56–57
Necturus, 183
Neilsen, Eigil, 68
Neoceratodus, 33, 74, 92, 162,
    215
Nerine, 22, 29
Nesides, 94
New York Aquarium, 229–30
New York Times, 52
ngombessa (gombessa), 55
Norman, J. R., 37, 40
notochord, 32, 157
    see also Latimeria, notochord and
    vertebral column
nyessa, 55
    see also Ruvettus

oil, in tissues, 32, 132, 139
    see also Latimeria, depth; lungs
Old Fourlegs (Smith), 52, 55, 67
opossum, *see Didelphis*
Origin of Species, The (Darwin), 71
osmosis, 178
Osteichthyes, 81, 180
ovoviviparity, 90, 193
    see also Latimeria, reproduction

paddlefish, *see Polyodon*
Pamanzi, 51, 53, 56, 67
    see also Comore Archipelago
Pangaea, 95
Peabody Museum of Natural
    History, *see* Yale University
Pemba, 47
Periophthalmus, 151, 163, 217
Pickford, Grace Evelyn, 177,
    182
placenta, 193–94
    see also reproduction, general
Polyodon, 75, 82
Polypterus, 82, 87
Pope, Alexander, 203
population size, *see Latimeria,*
    population size

Port Elizabeth, 20
poster, 47
preadaptation, 214
Prionace, 153
Protopterus, 75, 92
prototetrapod, 215
Pseudotriakis, 170

radio cocotier, 56
Rana, 183
Rand, Charles, 198
ray-finned fishes, *see* Actinopterygii
Reading (England) meeting, 203–04
rectal gland, 182
    see also Latimeria, physiology
Red Sea, 46, 99, 151
reefs, coral, 43, 104, 106, 146–47
relationships, evolutionary, 202–18
    see also Latimeria, relationships
reproduction, general, 90, 190–201
    see also Latimeria, reproduction
reptiles, 43
respiration, early tetrapods, 217
respiration, coelacanths, *see*
    Latimeria, respiration and
    metabolism
Réunion, 44, 101, 148–49
Réunion hot spot, 100
Rhabdoderma, 85, 89, 196, 199
Rhipidistia, 92, 180, 206
Rhodes University College, 22
Rift Valley, African, 95
Robineau, Daniel, 123, 221
Rodriguez, 101
rostral organ, 40, 92, 175
    see also Latimeria, brain and senses
Royal Society (London), 108,
    120–21
Royal Society of South Africa, 41
Ruvettus, 55, 107, 120, 136–39, 141,
    227, 232
Ruvettus, fishery, 137

Sarcopterygii, 82, 203, 212
Sargent, W. E., 25
scales, 42
    see also Latimeria, age
Schaeffer, Bobb, 68, 198

Schultze, Hans-Peter, 196
Scientific Council for Africa, 64,
    148
Scoones, Peter, 109
Sea Fishes of Southern Africa (Smith),
    46, 52
seafloor spreading, 95
Seychelles, 67, 99, 100, 149
shark, white-tipped, *see Carcharodon
    albimarginatus*
sharks, 120, 121, 153–54, 175,
    194
Simons, Elwyn, 116
Smith, J. L. B., 22, 26–69, 77, 117,
    123, 138–39, 145, 181, 189,
    223–24
Smith, J.L.B. Institute of
    Ichthyology, 69, 224
Smith, Margaret M., 22–29, 35,
    46–49, 53, 69
Smuts, Jan, 46, 51
snapper, *see Etelis*
Soha, 52, 55, 60
Somali Plate, 190
South African Fisheries
    Commission, 43
Squalus, 120, 153
Steinhardt Aquarium, 115
Stevens, Jacques, 108–09, 160
submersibles, 226–29
    see also Fricke, Hans
swim bladder, *see* lungs
swimming, general, 153–64
    see also Latimeria, fins and
    swimming

tail fin, 62
Tananarive, 48, 54, 65, 66

Tanganyika, 47
temperature, water, 142
    see also Latimeria, temperature
Tethys Ocean, 148
tetrapods, 164
    origin, 43, 202–18
Thomson, Keith S., 116, 166–69,
    172, 184–96
Thomson, Linda, 119
Times (London), 38
trimethylamine oxide (TMAO), 170

Undina, 31, 90
urea, 150–83, 197
    see also Latimeria, physiology

vertebral column, *see Latimeria*,
    notochord and vertebral
    column
viviparity, 90, 193
    see also Latimeria, reproduction

Watson, D. M. S., 33, 37, 90, 196,
    198
White, E. I., 37–39
William Scoresby, 46
Woodward, Arthur Smith, 30, 34,
    37, 42
Wourms, John, 201

Xenopus, 183

Yale specimen, *see Latimeria*
Yale University, 116, 122, 159
ylang-ylang, 106

Zanzibar, 47–49
Zelée Bank, 100, 104